MW00974576

African Cichlids I: Cichlids from West Africa

Horst Linke · Wolfgang Staeck

Horst Linke · Wolfgang Staeck

AFRICAN CICHLIDS I

CICHLIDS FROM WEST AFRICA

A Handbook for their
Identification, Care, and Breeding

The African Cichlids will be presented
in two volumes:
 I: Cichlids from West Africa
II: Cichlids from East Africa

Cover photographs:
Above left: *Hemichromis cristatus*
Above right: *Steatocranus casuarius*
Below left: *Tilapia buettikoferi*
Below right: *Pelvicachromis pulcher* "Red"
Rear: *Pelvicachromis taeniatus* "Lobe"

Photo insertions: Cichlid biotope between Calabar and
Mampfe in eastern Nigeria

© 1994
Tetra-Press
Tetra Werke Dr. rer nat. Ulrich Baensch GmbH
Herrenteich 78, D-49324 Melle, Germany
All rights reserved, incl. film, broadcasting,
television as well as the reprinting
1st completely revised edition
Printed in Germany

Distributed in U.S.A.by
Tetra Sales U.S.A.
3001 Commerce Street
Blacksburg, VA 24060

Distributed in UK by
Tetra Sales, Lambert Court,
Chestnut Avenue, Eastleigh Hampshire S05 3ZQ

WL-Code: 16755

ISBN 1-56465-166-5

INDEX

PREFACE

Our thanks are due to Professor Dr. DIRK THYS VAN DEN AUDENAERDE, Director of the Royal Museum for Central Africa in Tervuren as well as to Mr. OTTO GARTNER of Vienna, who has gathered much new information during his travels to West Africa.

Berlin / Schwarzenbach am Wald, March 1993   Horst Linke / Wolfgang Staeck

THE CICHLIDS

Cichlids range amongst the most popular fishes for the aquarium. They are often considered "teachable" inhabitants of the tanks. Their behaviour is very interesting, and they are correctly regarded as intelligent fishes favoured by many aquarists. The African species especially often have very conspicuous, even "iridescent" colours. Therefore, many associations all over the world have been founded whose members deal with the captive husbandry and breeding of Cichlids exclusively. Two groups of Cichlids are distinguished, i.e. large types and those which are adequate for keeping in tanks of moderate size. These "smaller" Cichlids, which do not exceed 12 cm in length, are usually appropriate for a keeping in richly planted aquaria. Especially noteworthy in this respect are the so-called Dwarf-cichlids which can also be kept in fairly small tanks.

Cichlids are often considered aggressive and irascible, but this statement is largely based on improper keeping of large-growing species in tanks which are too small. The

The Cichlids of Africa especially often have very conspicuous, sometimes almost "iridescent" colours as shown here in *Pelvicachromis taeniatus* Colour-morph "Muyuka" (top: ♀, bottom: ♂).

species introduced in this book belong almost exclusively to the medium- to small-sized forms. With only three exceptions Cichlids exclusively inhabit America, Africa, and Madagascar. During the past decades a flood of new discoveries swept over from the Black Continent. Many new species come from the large lakes in East Africa, and names of fishes from Lake Victoria, Tanganyika, and Malawi became buzz words for many aquarists. Malicious tongues talked about fashion in aquarium keeping which also had financial impacts. Nevertheless, during the past few years the situation turned back to normal. It was recalled that Cichlids not only come from East Africa but also from West Africa. Our books undertake the attempt to introduce a large portion of these species of fishes. For this purpose they were divided into a first volume which deals with West Africa Cichlids and a subsequent second one which focuses on those from East Africa. By way of introduction some basics should however be discussed.

Whilst Cichlids from East Africa mainly inhabit the large lakes and only a small number of species lives in the streams of the surrounding steppes, the Cichlids of West Africa usually occupy clear currents and rivers of the vast rainforest areas and only a small percentage of these species inhabits watercourses crossing the bushlands. The different behavioural patterns are noteworthy. Generally all Cichlids may be considered bottom-dwelling species. Only a negligible percentage can be regarded the exception.

Whilst the inhabitants of the East African lakes are predominantly mouthbrooders with a mother-family structure (i.e. the females care for and protect their offspring), West Africa has mainly other forms of parental care behaviour. A large portion of these fishes from rainforest areas belongs to the cave-spawning Cichlids. They usually form a "mother-father-family" during times of reproduction with the female clearly being the more active part. A relatively high number furthermore belongs to the substratum spawners in which both parents take care of the fry with equal intensity.

A very interesting breeding behaviour is found in the few West African mouthbrooders which also belong to the secretive breeders. They partly take care of the fry in true parent-families (i.e. both parents alternatingly carry the fry in the mouth at certain intervals) or in a "father-mother-family" where sometimes only the male hatches the young fish in his mouth. Thereafter, the care of the fry is shared by both partners and both provide shelter for the young fish in cases of danger.

This picture from the Lake Malawi shows typical algae grazers. They are very common.

Many Cichlids are mouth-breeders: *Tropheus duboisi* with eight day old fry.

Cichlids are however not only distinguished by behaviour, distribution, and breeding biology, but also by body shape and other features. A system for arranging the organisms in order is necessary for proper differentiation. This system was suggested by the Swedish botanist Carl VON LINNÉ in the year 1758. Despite of vast new knowledge gathered during the course of the centuries, this system is still in use. It arranges the variety of forms of plants and animals and also shows the systematic position of the Cichlids. In the class Osteichthys, the bone-fishes, one finds in the order Perciformes, the perchlike fishes, and the suborder Percoidea, the perch-fishes, the family of Cichlidae, the Cichlids which contains more than 100 genera. This view into the systematics may be considered unnecessary by most of the aquarists. In the 10th edition of his book "Systema naturae", LINNÉ combined the systematic order with the binary nomenclature. This means the double naming for the scientific arrangement. Even with this system some aquarists experience problems and frustrations due to the unfamiliar scientific designations.

Therefore the common names are very much in use which however differ and do not provide a clear and exact nomencalture. Only the scientific names allow an unmistakable identification worldwide. This may be illustrated by an example. Many aquarists have kept the mouthbreeder from West Africa which is often referred to as Gunther's Cichlid. Despite this widely used name, confusion often occurs. If the scientific name *Chromidotilapia guntheri* is however used, confusion is eliminated. For the informed aquarist this name however means much more. Staying with this example and writing it in full, one gets *Chromidotilapia guntheri guntheri* (SAUVAGE, 1882). The first name *Chromidotilapia* indicates the genus. It is always capitalized. One may recognize the relationship provided one is able to assign it to the family. The second name, *guntheri* in this case, refers to the species and is never capitalized. It has priority, which means that even if the generic name changes, the fish still keeps the species name introduced by the first describer. Since both generic and specific name are Latin or Latinized and the species name has

The most familiar aquarium-fishes of Lake Malawi inhabit the rocky shores.

to be of the same sex as the genus name, only the last syllable of the species name may change to fit the new generic name. The third name *"guntheri"* shows that there is another subspecies of this species. The fourth name (SAUVAGE, 1882) indicates that this fish was described by the French scientist SAUVAGE in the year 1882. Since the name is indicated in brackets, it is obvious that the generic name has been changed at a later date. SAUVAGE described this species in 1882 as *Hemichromis guntheri.*

These remarks on the correct names of our fishes may look somewhat complicated at the first glance, but clearly show that they are a necessity for an accurate correspondence whereas the various vernacular names may only serve as supporting terminology.

Finally there should be a brief look at the description and outer organisation of the Cichlid body. "Head" refers to the zone from the tip of the snout to the hind edge of the gill-cover, to the gill opening. The part from there to the anus is considered the body. The remaining part from there to the base of the tail where the caudal fin begins is referred to as the tail. The caudal peduncle on the other hand ranges from the hind edge of the anal-fin to the base of the tail. Cichlids have a dorsal fin on the back, a pectoral fin on either side of the anterior body, and two often extended ventral fins usually at the level of the pectoral fins. In the posterior part of the lower body there is an anal fin. At the end of the body the caudal peduncle carries the caudal fin. This has an impact on the measuring system applied. The total or over-all length, often referred to just as length, is measured from the tip of the snout to the end of the caudal fin. The body-length, sometimes also named standard length, indicates the distance from the tip of the snout to the end of the body, i.e. up to the end of the caudal peduncle. The caudal fin is thus not taken into consideration.

With these explanations the introductional part may be considered completed. The following first part of this work deals comprehensively with the Cichlid fish-fauna of West Africa. The Cichlids of East Africa will be introduced in the second volume.

The water-courses of West Africa are usually overgrown by the vast green of the rainforests.

THE MODERN AQUARIUM FOR CICHLIDS OF WEST AFRICA

All Cichlids appreciate much space for swimming. Since Cichlids are bottom-dwelling fishes, special attention should be paid to the available ground space in their aquaria. This is the main habitat for the fishes and therefore must not be too small.

Since this first volume of the work deals with species which, with a few exceptions, are tolerant towards plants, there is no need to do without a decorative arrangement of plants. Due to this the tank for Cichlids should not

fall short of a certain height. Lengths of 130 cm should be considered the minimum for Cichlid aquaria. In order to obtain an overview about the technical requirements, we would like to describe the procedure of setting up and decorating an aquarium for West African Cichlids.

First of all an aquarium needs to have a stand. Complete cupboards made of wood of choice and corresponding with the furniture of the lounge are popular. If one builds the stand oneself, one should obviously take into

There is no need to forego on an attractively rich planting in an aquarium for West African Cichlids.

consideration that it has to be sturdy and is carried by the floor without being exposed to vibration. An aquarium of 130 cm in length, a depth of 40 cm, and a height of 50 cm, including accessories, weighs about 300 kg (!) without the stand. The carrying capacity of the floor has thus to be checked. Furthermore one should be able to observe the action in the tank comfortably. The normal sitting position offers the best possibility. The stand has to be adjusted to a horizontal position by means of a spirit-level in order to obtain a regular level of water at the top edge of the tank. Furthermore the stand has to provide a smooth and straight surface for the tank to stand on. A thin sheet of styrofoam the same size as the tank is placed between the stand and the tank to provide a flexible buffer. Since styrofoam tends to deform under pressure, it is advisable to cut out 4 to 5 pieces of about 200 × 150 mm from the central parts of this sheet which prevents a possible material jam.

The illumination of the aquarium with natural light is not recommended because of the irregular and changing light value. Artificial illumination should therefore be given preference! If the tank stands free and is to provide a view into the underwater-world from above as well, the choice should be well-shaped mercury-power lamps which are set up hanging above it. If lighting tubes are given preference, the available lengths should be taken into consideration when the tank is purchased. Tubes of 18 W are 59 cm in length, i.e. measure about 65 cm with the fittings and thus are appropriate for a tank of 70 cm. 36 W lighting tubes measure 120 cm in length with another 6 cm for the fittings and thus fit a 130 cm aquarium. A tube of 58 W has a length of 150 cm to which the fitting must be added thus consuming 156 cm and being ideal for a tank of 160 cm.

Each aquarium should have a rear wall. This provides a feeling of better security to the fish and adds to the over-all appearance of a well-decorated tank. It is just not attractive if a view into the underwater landscape includes the flower-patterned wallpaper behind it. Various options are available. Pet-shops have a variety of rear walls in the form of photographic posters which are mounted from the outside onto the rear side of the tank. Dark coloured decoration cardboard in a is another option. The ideal choice however is an irregularly shaped rear wall with small cavities which is placed in the aquarium and thus forms part of the underwater world. Various methods have been tried using materials such as cement, wood, cork, and similar elements. The easiest and best value-for-money is styrofoam which unfortunately is rarely found in the range of products offered by the pet-shop business. Even less skilled D.I.Y.'s will however be able to manufacture such walls without problems.

The production can be briefly described as follows. One acquires styrofoam sheets which are usually offered at sizes of 50 × 100 cm with a thickness of 10 mm or thicker. For our rear wall we utilize those of the 50 mm type. To model them a soldering-iron of 80 W is best suitable. The copper tip of the soldering-iron is filed to a knife-shape. With the hot iron the hard styrofoam can be cut like butter. The straight tip is used to cut the sheets to size and it is determined which side will become the front and which the rear. The edges of the future rear side are cut off to allow a perfect fitting in the tank and to prevent them pressing on the rounded inner corners of the aquarium. Two plates are to be fitted if the size of the tank exceeds 1 metre. Two sheets are therefore cut to equal size so that they meet below the often present central flange of the aquarium.

An accurate fitting of the styrofoam rear wall is of great importance since the sheets tend to drift up due to their inherent buoyancy. To shape the front side of the sheets, one bends the tip of the soldering-iron with tongs to a hook-shape and cuts appropriate "sausages" out of the styrofoam. There are no limits which restrict the fantasy. Small caves and cavities which may be filled with sand and small plants later greatly contribute to the over-all appearance. The thickness of the material provides various possibilities.

Once the rear wall is finished, it looks rough and edgy. In order to smooth this, the styrofoam sheets are held over a burning candle or a lighter with a moderate flame. One allows the styrofoam to flow until the surfaces and edges are rounded. Now the rear wall is completed and is due to receive its painting. Unthinned dispersion colours which are usually used for colour-shading of wall-paints can be obtained from any paint-shop. Dark colours like brown, dark grey, or black are recommended, but each person is free to creatively mix colours according to taste. The paint is applied with a paint brush and all sides, including the rear side, are painted. Attention should be paid to the fact that the paint cannot be removed easily from clothes and other materials once it is dry. It is soluble in water initially and all tools used can be cleaned easily while the paint is wet.

After the painting, allow the rear wall to dry at normal room temperature for two or three days, then paint once more if necessary and let it dry again. Before it is placed in the aquarium it should be "soaked" for a couple of hours in a water-filled basin, e.g. a bath, and then rinsed with fresh water. Styrofoam rear walls of this kind last for many years and have proven successful in aquaria even with sensitive fresh- and sea-water fishes.

Now the rear wall is placed in the aquarium and special attention is paid to its correct fitting. Thereafter the soil is added. Fine dark gravel with a grain-size between one and three millimetres is recommendable. Larger sizes, e.g. of pea or bean size, should not be chosen since they serve primarily to collect dirt. I have not come across a biotope in Africa which had provided such coarse gravel. Unless contaminated with other substances, the unwashed fine dark gravel is filled into the tank up to a height of 1 cm. If the substrate contains some loam it is of advantage for the growth of the plants.

If the choice is not to use a heating-rod as source of heat, heatingwire provides an alternative. The wire is switched on for a short time until it becomes hand-warm and is then laid out in spirals on a smoothed layer of gravel. Its capacity should be approximately 0,2 to 0,3 W per litre of water which needs to be heated. Another 1 cm layer of gravel is used to cover the heating-wire. In order to provide the plants with nutritional substrate, a thin layer of a modern nutrient stabilizer consisting of iron and organic substances is placed in the aqua-

rium. The additive corresponds to tropical soils and should be prepared one day before being used. The mud-like substance is mixed with some gravel and then applied all over the ground. This layer is eventually covered by another 3 to 4 cm layer of gravel which preferably should be somewhat higher in the rear part.

After this, decorations such as pieces of bog-oak and stones can be arranged in the aquarium. As far as stones are concerned one should ensure that they are calcium-free which can be tested by dropping some hydrochloric acid on them. If the surface starts foaming and dissolves, the stones contain calcium and are therefore unsuitable for the aquarium. Decorations should be placed directly onto the bottom plate. Through digging activities many nice caves have collapsed and killed fish or even damaged aquarium glass. In addition to wood and stones, coconut shells with a small opening are frequently used since they blend into the underwater landscape very well.

Upon completion of the decorating and the erection of small rockcaves or stone formations the water may now be added. In order to avoid the jet of water to stir up the soil, the substrate is covered with paper and a shallow container or pot is placed onto the paper. The water is then carefully poured into the container until the tank is half filled. Unfortunately it will mostly be the ordinary water from the tap which cannot be considered ideal for the purpose. Usually medium to very hard water runs out of the piping systems of the local waterworks, i.e. water with a high mineral content. Accordingly the reaction of this water is at an alkaline level, and thus is a substance very alien to our fishes in their natural habitats. Everybody should therefore establish the qualities of the water by own measurements.

If clean rainwater is available, one may come closer to natural water by mixing it with the municipal water. It should however be stated that pure rainwater or very soft water deficient in minerals is often too instable for the normal keeping, but ideal for breeding purposes. My experiences have shown that the best values for the normal keeping range between a total and carbonate hardness of 5 and 7 °dH with a constant pH of 6,5 to 6,7.

With the tank half filled with water, one can now begin to arrange the plants. The background should obviously be furnished with highgrowing stem-plants which are arranged in larger groups and planted in circles a centimetre or two apart. A sparse planting here would be saving in the wrong place! Only a richly vegetated aquarium guarantees an attractive look and furthermore provides cover for the fish. As Aquatic plants also play an important role in the biological maintenance of the water, at least 20 stem-plants of each species should be arranged in groups. The various plant groups should contrast from one another in the shape of the leaves and colour. Soft green pinnate leaves of *Cabomba caroliniana* plants may be placed next to the reddish brown leaves of *Rotala macranda,* whereas the elongate narrow green leaves of *Hygrophila stricta* look good in contrast to the Red Milfoil *Myriophyllum* "matogrossense". The individual groups may be arranged separately from the rear wall up to the centre of the aquarium, and the sides may also be planted with stem-plants growing up to the water surface. Spare areas of the rear wall can be planted with the Javanese fern *Microsorium pteropus* which may be attached with glass pins. For the central ground zones, small bushy plants are the best choice. Red and Green Tigerlotus, *Nymphaea lotus,* should not be ignored. Initially growing as a round bush with large leaves, it develops floating leaves below the water surface after some time providing preferred places for the fish to stay. When choosing plants one should pay attention to the potential of rapid growth so that the nutrients are consumed quickly by the plants and the growth of algae in a new aquarium is prevented from the beginning. It will be several weeks before the aquarium runs properly and the plants present themselves in all their rich array after which more complicated and slower-growing plants should be added. The actual planting scheme is however left up to individual taste. Once the planting is completed, the water level is carefully brought up to maximum. Thereafter a circulation pump with a filter-pot is installed in one of the rear corners of the tank. For an aquarium of 130 cm in length the circulation pump should have a capacity of 600 litres per hour. The filter-pot need only have a mechanical filter substrate. It is important to install the water outlet horizontally 2 to 3 cm below the water surface. The resulting current moves the water surface rapidly and splashing and gushing should be avoided. The current itself should be directed from the rear corner of the aquarium to the first third or the centre of the front glass plate in order to provide in a relatively good water circulation. Experience has shown that it is favourable to have the water sucked up in the same corner where it is expelled. The desirable side-effect of this water movement is that almost all decay, which cannot settle in the fine gravel, is drawn into the filter-pot by the current. A further advantage is that the plants are constantly furnished with nutrients. By "ripping" the water surface open additional oxygenation of the water takes place. The filter-pot of the pump is covered with some plants and thus invisible to the observer. External filters which circulate the water through hoses from and to the aquarium are not ideal for a successful movement of the water. The temperature of the water can be regulated automatically by a thermostat which is hung into the water. An even better solution is however to utilize modern technology and bring the temperature device out of the tank. Electronic thermostats are easy to operate and allow the regulation from outside. A small waterproof-sealed probe regulates the chosen temperature. This aquarium, set up according to modern technology, now still needs illumination required for the growth of the plants.

The two basic alternatives were referred to already in the introduction. For the usage of lighting tubes the general formula 0,5 W per litre of water applies. A new generation of lighting tubes, i.e. the Indium-Amalgam-technology introduced by Osram, allows to undercut this formula to some extent. These Lumilux lighting tubes have a 70% higher light-production than the commonly used Osram de Luxe types. For the illumination of an aquarium the light-colour LUMILUX 21, and if several tubes are required, a combination of LUMILUX 11 and 21 are recommended. The full extent of the additional light is however only gained if the tubes are exposed to temperatures between 30 and 40 °C thus making it necessary to leave the lid open on top since the temperatures rise much higher in closed constructions despite ventilation openings. Regardless of the new lighting-tube technology, the exchange of worn tubes still remains a necessity.

After running a ordinary lighting tube for only half a year its light intensity has already reduced by a third. The colour of the light has changed thus having a negative effect on the growth of the plants. This necessitates the replacement of the tubes at regular intervals. If tube-assemblies are used, these should not all be replaced at once, but in sequence with longer breaks in between.

In the tropical home of the Cichlids a day is followed by a night of almost the same duration which means that a day lasts 12 hours and so does the night. This rhythm should accordingly be followed in the aquarium as well. The illumination is therefore in use for 12 hours every day without exception, no matter whether workday or holiday or whether the keeper is at home or not. In order to have the day starting always at the same time, the use of a timer is strongly recommended. It is therefore of secondary interest when the illumination begins, it is only important that it ends after 12 hours, and that this happens every day at the same time.

A last "technical accessory" requiring attention is the nutrient carbon for the plants. Occurring naturally in the wild, carbon dioxide has to be added to the water artificially and transported to the plants. The addition of carbon which is absorbed by the plants only as the gas CO_2 and which leads to healthy growing plants is not, as many people may think, a "modern day fad" and should not be ignored as such. The statement that aquarists of the old days did not know about these things and also had a good plant-growth is simply false. The old foggies of yesteryear, about 40 years ago, were closer to the topic "carbon dioxide as fertilizer in the aquarium" than we are willing to recognize today. An example of this is the result of research undertaken by Mr. K. HEINRICH of the aquaristic society "Seerose" of Berlin-Lichtenberg, Germany, which was published in a weekly magazine in 1938. In this paper, the author points out that fish display more intense colours and a better appetite after carbonic acid was added to the water, the reason being probably the fact that the water was slightly acidified. However, the statement that the plants drastically changed thereafter is more important. *Ambulia* increased the diameters of their stems from 0,8 to 1,9 mm. *Aponogeton ulvaceus* changed their leaf-sizes from 23×128 to 45×200 mm, and *Cryptocoryne griffithii* increased theirs from 35×65 to 60×128 mm. A young plant of *Ceratopteris thalictroides* grew from 20 to 270 mm in only eight weeks.

This report shows that the importance of the addition of CO_2 for the plants was already recognized at this time. The fertilization may be achieved by various methods.

Tests with the so-called "baking yeast bottle" were quickly discarded as its usage was too messy and indirectly contributed to the contamination of the water. The direct supplementation of CO_2 through a discharger is very dangerous and may easily lead to the poisoning of the fishes if an incorrect dosage is applied. Depending on the contents of calcium, different amounts of carbonic acid are absorbed by the water and thus the supply of CO_2 through a diffusor is the easiest and safest way. A strong and healthy growth of the plants will justify the effort. Caring for the plants means caring for the water and eventually results in healthy and active aquarium inhabitants.

Since we are faced with different behavioural patterns in the West Africa Cichlids, one should take their various instinctive actions into consideration when planting the aquarium. The smaller species of the genera *Nanochromis* and *Pelvicachromis* do not damage plants. However, as is the case in all Cichlids, some "minor adjustments" in front of the breeding cave take place.

As species of the genus *Chromidotilapia* often search for food by chewing the soil, it is possible that very small foreground plants, such as *Elatine macropoda* or *Marsilea crenata* are loosened out of the ground. Only certain species of *Hemichromis* require areas of unhindered sight during the reproduction period for the protection of their fry. These species are then no longer suitable for richly planted tanks. The large-growing Tilapias are also not advisable and the few species portrayed in this book should therefore be considered exceptions.

The tank for West African Cichlids is now completed. One should however wait for another fortnight since the aquarium should first stabilize. The chosen temperature should be monitored by a thermometer and all other technical appliances should also be checked during the next days. After about two weeks one will find that the plants have rooted and the first growth will be established.

Biotope in the Kasewe Forest in Sierra Leone

CICHLIDS FROM WEST AFRICA

It is no easy task to define the borders of the so-called West Africa. Generally the term is applied to the countries from Senegal in the northwest up to Nigeria in the Bay of Biafra. Literally and actually it is the area of the Upper and Northern Guinea Ledge extending from the mouth of the River Senegal in the west to the Cameroon Mountains in the east. In a wider sense one however includes the coun-tries south of the Cameroon Mountains which are considered to form the Lower or South Guinea Ledge. The connecting coasts of the Gulf of Guinea including the Bay of Biafra play a role here. Originally a coastal fringe of rainforest some hundred kilometres wide ranged from Senegal through Cameroon up to the central interior of the Congo Basin. Parts of this landscape has changed drastically in

Chromidotilapia linkei is a colourful discovery from West Africa.

THE FISH REGIONS OF AFRICA

(after POLL 1957)

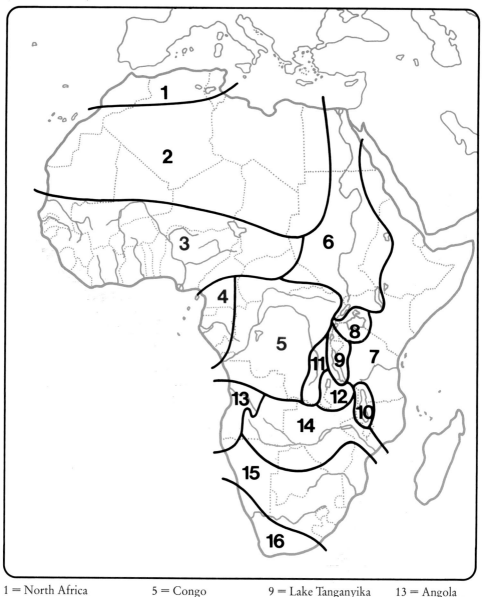

1 = North Africa	5 = Congo	9 = Lake Tanganyika	13 = Angola
2 = Sahara	6 = Nile	10 = Lake Malawi	14 = Zambezi
3 = West Africa	7 = East Africa	11 = Lualaba	15 = South Africa
4 = Cameroon and Gaboon	8 = Lake Victoria	12 = Luapula	16 = Cape

The first volume covers the regions 3 to 5, the second one 7 to 10.

recent times. In the areas of Togo and the Republic of Benin, the coastal forest zones have changed into dense bushlands not at least by the influence of man.

West Africa from the northwest to the south thus includes the countries of Senegal, Guinea-Bissau, Guinea, Sierra Leone, Liberia, Ivory Coast, Ghana, Togo, Republic of Benin, Nigeria, Cameroon, Rio Muni, Gabon, and Congo. As to how far one may consider Zaïre a West African country is doubtful since it has access to the west coast only by a very narrow strip of land and is rather a centrally situated state, but it may be included here.

Besides the rainforest, savannas are another major type of landscape in West Africa. It includes many so-called "small" rivers which

source in the North or South Guinea Ledge. After a short course they flow into the Atlantic Ocean. They are however only small in comparison with the few large rivers of West Africa. This means, they often have a length of 500 km, but look modest if compared with the approximately 1500 km of the Rivers Volta and Senegal, the Niger with almost 4000 km, or the Congo with some 4400 km.

There is ample water available all over West Africa. Daily showers, even during the so-called dry season, are responsible for a rich water supply in the rainforest areas. Myriads of various small water courses cross the landscape creating biotopes for an interesting and colourful underwater fauna. As far back as 150 years ago the first fishes from these areas became

Being a larger growing Cichlid, *Tilapia mariae* is adequate only for very large aquaria.

known. The last decade of the previous century saw the description of numerous species and of special interest was the genus *Pelmatochromis* as can be found in the "Catalogue of the Freshwater Fishes of Africa". Already by 1915, approximately 38 species were listed of which 12 originated from East Africa. Amongst them were some of the most beautiful aquarium fishes. During the following years various species were removed from this genus until some 50 years later a preliminary revision by the ichthyologist DIRK THYS VAN DEN AUDENAERDE reduced this genus to a few species. Old generic names were revived and new ones added thus enriching the vocabulary of the aquarists from then on.

Beginning with a paper titled "A preliminary contribution to a systematic revision of the Genus *Pelmatochromis* HUBRECHT sensu lato (Pisces, Cichlidae)" in 1968, a number of alterations were initiated and a process of thinking for all those interested in ichthyology. After this work only a few species were left in the genus *Pelmatochromis* and the majority were transferred to other genera. The formulation "once there were some *Pelmatochromis*" brought it to the point.

The paper by Prof. THYS implicated that names such as *Pelmatochromis camerounensis*, *P. aurocephalus*, *P. kribensis*, and *P. klugei* were to be deleted from the aquaristic vocabulary. Whilst *Pelmatochromis camerounensis* and *P. aurocephalus* were assigned to the species *Pelmatochromis pulcher* and *P. sp. aff. pulcher* respectively, *Pelmatochromis klugei*, *P. kribensis klugei*, and *P. kribensis* were recognized as *Pelvicachromis taeniatus*. Taxa such as *Pelmatochromis arnoldi* and *P. annectens* were consolidated in *Pelmatochromis ansorgii*. This resulted in some incorrect or confusing scientific or trade names to receive a valuable adjustment. Dr. THYS had worked out the revision and differentiation.

In the same year, a second paper appeared in which Dr. THYS presented a list of species belonging to the genus *Tilapia* (THYS 1968). Here, species such as *Pelmatochromis buettikoferi*, *P. ocellifer*, and *P. congicus* of the subgenus *Pelmatochromis* were listed as *Tilapia* and thus removed from the genus *Pelmato-*

chromis. Also *P. ruweti* was referred to as *Tilapia ruweti* now. Thus only *Pelmatochromis corbali* was left in the subgenus *Pelmatochromis*. The 1971 paper by THYS and LOISELLE titled "Description of two new small African Cichlids" eventually resulted in the elevation of the subgenera into generic ranks.

Another year later one more paper was published dealing with West African Cichlids. The authors were LOISELLE and WELCOME (1972) in this case. Based on old and new collection material they suggested a new genus in honour of Dr. THYS, i. e. *Thysia* LOISELLE & WELCOME, 1972. *P. ansorgii* was designated as the type species and was thus named *Thysia ansorgii* now.

One year later, in 1973, another adjustment became effective. Dr. E. TREWAVAS of London published on the systematics of Cichlids reintroducing *Pelmatochromis* as a genus. The genus contained the three species *Pelmatochromis buettikoferi*, *P. ocellifer*, and *P. nigrofasciatus*. Thus, two species have been re-transferred. A new genus was suggested by her for *Pelmatochromis congicus*. The new name was *Pterochromis* and the species therefore bears the name *Pterochromis congicus* today. A re-examination of the 1933 material of *Chromidotilapia exsul* furthermore revealed that this was the juvenile form of *Hemichromis bimaculatus* GILL and the taxon was therefore deleted. This species is no longer valid.

However, in order to not only follow the systematic changes of the original genus *Pelmatochromis*, a look at the other West African Cichlids is necessary. In 1976, Dr. TREWAVAS in co-operation with the English scientist A.I. PAYNE described a new species of *Hemichromis* as *Hemichromis fugax*, but it soon turned out that this new form was in fact the existing species *Hemichromis bimaculatus*. A second surprise happened in the same year. With the paper "An Ecological and Systematic Survey of Fishes in the Rapids of the Lower Zaïre or Congo River" by TYSON R. ROBERTS and DONALD J. STEWART several new species of the genus *Nanochromis* became known, amongst which was an "old fellow" which was erroneously taken for as *Nanochromis nudi-*

ceps so far. Described only on this occasion, it is to be referred to as *Nanochromis parilus* now. A preliminary finalisation was formed by the work of Paul LOISELLE "A Revision of the Genus *Hemichromis* PETERS, 1858)" and the 1987 revision by P.H. GREENWOOD "The genera of pelmatochromine fishes (Teleostei; Cichlidae): A phylogenetic Review".

Numerous new species were only described in the past few years and also imported alive.

Honour for this is due to the so-called "travelling aquarists" as they are sometimes referred to as. Often they were not shy of efforts and costs to shed some light on the darkness and to compile information on natural habitats for the aquaristic. The new discoveries are much appreciated and they will not be last ones made. West Africa is still large and full of mysteries. But please allow us to present to you now what is known.

Chromidotilapia g. guntheri Colour-morph "Bosumtwi" (top: ♂, bottom" ♀)

23

It was only in 1985 that on the base of the paper by GREENWOOD the last small species of the originally large genus *Pelmatochromis* was transferred into the new genus

Anomalochromis

an operation which was requested by the specialists for a long time. The genus is presently monotypic.

◗ *Anomalochromis thomasi*
(BOULENGER, 1916)

is the only known representative and is often encountered in aquaria. A relationship with *Thysochromis ansorgii* was presumed on several occasions. *A. thomasi* appears to rather be related to the genus *Hemichromis*, the so-called "Jewel Fishes". Unfortunately this aspect was not given attention in the 1979 revision of this genus.

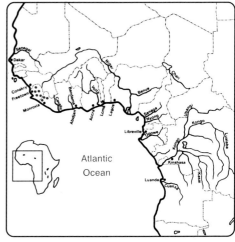

Distribution of *Anomalochromis thomasi*

The fish shown here originate from the Kasewe Forest in Sierra Leone. According to observations this species changes colour with

Anomalochromis thomasi ♀

the more south it occurs. In the southernmost parts of the country and in neighbouring Liberia, the animals are coloured more brilliantly. Both colour-morphs share the light blue small spots which run in horizontal rows all over the body.

Distinguishing the sexes is not always an easy task. Large males may reach 8 cm in total length whereas the females only grow up to 6,5 cm. The spotted pattern is somewhat more inconspicuous in the males so that they remain a little lighter than the females. Especially during courtship and spawning they display five more or less uninterrupted broad dark transversal bands and a dark back instead of the spots on body and caudal peduncle.

The

natural habitat

is in Sierra Leone and Liberia. It consists of small water courses crossing forested areas and bushlands. These are mainly very clean soft acidic waters rich in oxygen. This species shares its biotope with *Hemichromis bimaculatus*, *H. paynei*, *Pelvicachromis humilis* and others. They are relatively abundant in their habitats.

Care

Small aquaria may be sufficient to keep this species. Tanks of 130 cm in length and 50 cm in depth, richly vegetated and providing many hiding places in the form of caves and rock constructions are however more suitable. Bogoak also is a very good item for an attractive decoration. Richly planted grounds with interspersed lightings and flat smooth stones are especially appreciated by these fish. The species is very timid and tolerant towards plants. Soft, slightly acidic, and gently flowing water should be used for a healthy environment. The temperatures may range around 26 °C. Regular exchanges of water, i.e. a quarter to a third every week or fortnight, should be obligatory. Every type of living or flake food is suitable. Provided with these conditions, young fish grow rapidly and reach maturity after approximately 6 months.

For the

breeding

of this species one may use the same tank or choose another smaller one which is similarly decorated. Smooth, flat, rounded stones on a fine substrate of up to 2 mm grain-size are important. A few socializing fishes should also be kept in the breeding aquarium. When a pair has bonded and is ready to spawn, both partners display a brilliant pattern of transversal bands with the female appearing more contrasting than the male. After a thorough cleaning of the spawning site, the female lays approximately 300 eggs. Since this is a substrate spawner and the fry is taken care of by a parental family, both partners alternatingly care for the eggs. One parent always swims closely above the eggs and moves the water around them by the pectoral fins whilst the other partner guards the spawning territory and defends it if necessary. At a water temperature of 26 °C the embryos hatch after 48 hours and the male brings them to small depressions he has prepared before. The parental care often ends here and the larvae — but also very often already the eggs — are eaten by both parents. This procedure may be repeated five or six times since the pair would spawn again a short time later. Only when the hatchlings begin to swim in a school led by the parents, should newly hatched nauplii of the Brine Shrimp *Artemia salina* and powder food be offered. Especially in this species it can be observed that in the absence of a male, the female lays eggs by herself and guards them until they are spoiled. When feeding daphnia or tubifex caring for this prey can be observed as an act substituting parental care.

25

The interesting species of the genus

Chromidotilapia

are very different regarding their breeding behaviours.

The generic name was introduced by the British ichthyologist G.A. Boulenger in 1898, but dumped shortly thereafter until Prof. Thys eventually revived it after 70 years. This genus contains small to medium-sized Cichlids. They are vertically compressed with an increased height of the body. Their heads are generally large. As far as known up to today, all species are mouthbrooders with very different strategies. Another character for the species of this genus is the fact that both sexes have long pointed ventral fins. A warty, shoe-shaped eversion on either side of the roof of the pharynx is well developed. The pharyngeal teeth are regularly webbed and micro pharyngeal teeth are absent. The sexual dimorphism is very distinct in the majority of species. The females grow a little smaller, display a red belly in the courtship and spawning colouration, and have a metallic sulphur- to chrome-yellow, in some species even a bright light orange coloured longitudinal band in the dorsal fin. 12 species are presently recognized with partly different colour-morphs. The type species is *Chromidotilapia kingsleyae.*

1. *Chromidotilapia batesii*
2. *Chromidotilapia cavalliensis*
3. *Chromidotilapia finleyi*
4. *Chromidotilapia guntheri guntheri*
5. *Chromidotilapia guntheri loennbergi*
6. *Chromidotilapia kingsleyae*
7. *Chromidotilapia linkei*
8. *Chromidotilapia schoutedeni*
9. *Chromidotilapia* sp. "Shiloango"
10. *Chromidotilapia* sp. "Mondemba"
11. *Chromidotilapia* sp. "Atogafina"
12. *Chromidotilapia* sp.

Biotope near Calabar in southeastern Nigeria

▶ *Chromidotilapia batesii*

(BOULENGER, 1901)

Colour-morph "Kienke"

Distribution of *Chromidotilapia batesii* Colour-morph "Kienke"

This is a species — as will be described in this book in many cases which occurs in two colour-morphs. The morph "Kienke", named after the Kienke River-system in southern Cameroon is portrayed here first. The male reach a length of approximately 12 cm with the females staying only insignificantly smaller. Distinguishing the sexes is not always easy in this form. The metallic longitudinal band in the dorsal fin — in almost all species a definite trait for the determination of the sex — is rarely or even not at all recognizable. When reaching maturity, the females are somewhat more rotund in girth when compared with a male. During courtship seasons their belly regions are coloured wine-red below the brown transversal bands. This species has a relatively wide distribution.

Chromidotilapia batesii Colour-morph "Kienke" ♀

27

In the reference material, the

natural habitat

is indicated as the River Benito in Equatorial Guinea, also referred to as Spanish Guinea, and correctly quoted as Rio Muni after it became independent.

The species is widely distributed in this small country and often encountered here as well as in the neighbouring southwest of Cameroon where it inhabits the Kienke River-system.

These are rainforest rivers of different sizes. Mostly they are small, i.e. only 50 cm deep, and up to 4 metres wide. The streams are of a light brown colour with a constant temperature of around 25 °C which permanently flow, making their ways through the rainforests. Hardly any sunshine reaches the water surface through the dense foliage of the trees.

Table 1

Location:	Road from Kribi to Ebolowa, first stream after the exit of the village; southwestern Cameroon
Clarity:	very clear
Colour:	brownish
pH:	6,3
Total hardness:	below 0,5 °dH
Carbonate hardness:	below 0,5 °dH
Conductivity:	10 micro-Siemens at 25 °C
Nitrite:	0,00 mg/l
Depth:	up to 50 cm
Current:	slight
Temperature:	25 °C
Date:	15.1.1977
Time:	16.00 hrs

Branchwork and roots of trees form the favoured places for these fish. Wash-outs in the banks, partly submerged tree-trunks lying in the water, or areas of aquatic plants covering a couple of square metres are also utilized. The substrate consists of fine gravel with a grain-size of approximately 1 mm in diameter. The water flows constantly. Measurements taken revealed that there is an average current of 6 metres per minute, which means that every ten seconds the water of one metre in length is exchanged completely. An oxygen content of 7,0 mg/l was established.

The blooms of the *Crinum natans* are especially common during the month of February.

◗ *Chromidotilapia batesii*
Colour-morph "Eseka"

This apparently endemic form of the vicinity of Eseka differs from the colour-morph "Kienke" by numerous traits in colouration. Whilst the males grow up 12 cm, the females remain some 2 cm smaller. Besides the pattern of 4 or 5 transversal bands on the body, mature males display numerous small dark spots on the fin-edges forming a pattern of rows in the areas of the soft rays of the dorsal, anal, and caudal fins. The females lack this spotted pattern, but instead have a metallic chrome-yellow band in the dorsal fin. Their banded pattern on the body is feebly developed, but in the courtship and spawning colouration their bellies turn yellowish with the ventral part becoming faintly to deep wine-red corresponding to the tail.

Distribution of *Chromidotilapia batesii* Colour-morph "Eseka"

Chromidotilapia batesii Colour-morph "Eseka" ♂

29

The

natural habitat

of this colour-morph also lies in southwestern Cameroon. It consists of small water-courses in the northern catchment of the Nyong River in the area of Eseka.

Usually these are rainforest regions, rarely interspersed by plantations. Sometimes the water-courses are only 2 metres wide. The substrate consists of fine gravel of about 1 mm grain-size. Many rocks of different size are sometimes found. The majority of aquatic plants belongs to *Nymphaea lotus*, the Green Tigerlotus.

Care

Spacious aquaria are needed for this species. A minimum length of 130 cm and a depth of 50 cm must be provided. Fine gravel with a grain-size of up to 2 mm is suitable as substrate. Calcium-free rocks should be placed directly onto the base plate and arranged to form caves. Attention should be given to the fact that the ceiling of the cave is set up horizontally since this is the ideal place for spawning. Heating from the bottom is to be given preference. Further decoration may include various kinds of wood which are also placed on the base plate and arranged vertically. Only then should the gravel be filled in and also poured over the rocks and the wood, covering it partially. This may be unwashed but "clean" gravel. Only the top layer of approximately 1 cm should be washed.

Decorating an aquarium similar to the natural habitat requires emphasis on West African plants. This includes the many sturdy species of the genus *Anubias*, e.g. *Anubias afzelii, A. barteri,* and *A. gilletii* for the back-

Chromidotilapia batesii Colour-morph "Eseka" ♀

ground. The originally installed styrofoam rear-wall may receive some *Anubias*-species fixed with glass-pins or other aids in order to create an attractive upper part of the tank. They are also suitable for the front section.

Bolbitis heudelotii, the Congo-Waterfern, may also be attached to the rear-wall or planted in the foreground. One should ensure that only the roots are covered by the gravel and that the rhizome stands free. Anchoring down the rhizome, clamping in between rocks, or using fishing line to tie it on to a piece of wood is often sufficient. *Nymphaea lotus,* i.e. Green and Red Tigerlotus, produces a colourful atmosphere. Free space should be provided in front of the caveconstructions since *Chromidotilapia batesii* loves to dig out the spilled caves and needs space for this.

Soft water poor in calcium is preferred, but husbandry is also possible in so-called hard water. It should always be clean and constantly be in slight movement. Notwithstanding this, a regular exchange of a quarter to a third of the water every week or two remains a necessity. Temperatures may range around 25 °C and the illumination is to be switched on without interruption for 12 hours a day. Nutrious live food should alternate with flake food. Larger tanks may house one or two West African Cichlid pairs to provide company. However, several fishes giving life to the central and top levels of the swimming space are necessary, e.g. larger Gouramis or Tetras. This composition, with not too many inhabitants and a very dense vegetation, makes *Chromidotilapia* interesting objects to study.

Here, the

reproduction

plays a great role. With regard to the parental care, this species is distinctly different from other species. *Chromidotilapia batesii* belongs to the larvophile mouthbrooders which means that the females usually attach their eggs to the ceiling of a cave where there are instantly fertilized by the male. After the deposition of the eggs, the female cares for the clutch whilst the male more or less intensely defends the spawning territory. At a temperature of 25 °C the larvae hatch after approximately 70 hours and are then taken into the mouth by the female where they stay until they are able to swim freely. Only then the male again joins to care for the young fish. Larvophile mouthbrooding thus means that the parents, the female in this case, do not take the eggs in the mouth once they are fertilized as ovophile mouthbrooders do, but only take in the newly hatched larvae after approximately three days. Once the young fish swim free one should provide them with nauplii of Brine Shrimp. Powdered flake-food is included in their diet as well. The young fish are cared for by the parents for several weeks. It is important that a few other fish are in the same aquarium since this causes the parents to be especially wary and alert.

Table 2

Location:	Water-course on the road to Eseka. 50 metres after the turn-off from the road Edea-Jounde in southwestern Cameroon.
Clarity:	clear
Colour:	none
pH:	6,6
Total hardness:	below 1 °dH
Carbonate hardness:	below 1 °dH
Conductivity:	20 micro-Siemens at 23,5 °C
Nitrite:	0,00 mg/l
Depth:	up to 80 cm
Current:	little
Temperature:	23,5 °C
Date:	15.2.1979
Time:	09.30 hrs

◆ Chromidotilapia cavalliensis

(THYS & LOISELLE, 1971)

Although several preserved specimens were available for some time, the first living fish of this species were first imported only in 1988 by JÖRG and JÜRGEN FREYHOF and ANTON LAMBOJ. Because of their preparing support on site it was possible for one of the authors to record these fish in their natural biotope and bring some specimens back to the aquaria at home only a few weeks later. *C. cavalliensis* attains a length of approximately 12 cm in the males with the females growing only slightly smaller.

They show many parallels to *Chromidotilapia g. guntheri* as far as colour and pattern are concerned except that their colouration is much less brilliant. Like *C. g. guntheri*, *C. cavalliensis* is a mouthbreeder in the male sex underlining the parallels to the genus *Chromidotilapia*. This is one reason why this species did not remain in the genus *Limbochromis*.

Atlantic Ocean

Distribution of *Chromidotilapia cavalliensis*

The

natural habitat

of this species is situated near the village of Sahibly in the area of Toulepleu on the Cavally River, the frontier river between the countries

Chromidotilapia cavalliensis ♀

Ivory Coast (Côte d'Ivoire) and Liberia. There the fish inhabit the central and upper parts of the river.

Sahibly and the area of the Cavally River are still somewhat off the the beaten track. For better comprehension some notes about the journey to the biotope and the catching of the largely unknown fish are given here. Starting point of our trip was the small town of Guiglo in the west of Ivory Coast. This is a town with a couple of African hotels approximately 30 km west of Duekou). A good tar-road extends up to Guiglo, but ends right behind this town. From there a broad sand-track leads to Toulepleu 114 km farther. This laterite-red coloured road crosses hilly forest-areas, a region with a still fairly dense vegetation. It is 107 km to the Cavally River and it took us almost two hours to get there even though our driver rode at speeds between 60 and 80 km/h. The Cavally is crossed by a 60 metre long narrow bridge set up on high poles. The village Sahibly lies behind this bridge. The Cavally River runs approximately 10 m below the bridge in a bed of approximately 25 m in breadth. Its banks are vegetated with brush and trees. At low-water the depth is usually only 50 cm on the banks and up to 2 m in the middle at some places. There are some shallow sand-banks, and many trees and branchwork lie in the water.

The Cavally River runs slowly and may rise by one and a half metres during the rainy seasons. At the time of examination the water was clear to slightly murky and of light greenish colour. At 41 micro-Siemens at 27,7 °C the water was very soft (Total hardness below 1 °dH, Carbonate hardness 1 °dH). It had a pH of 6,7. The ground of the river was sandy and leaf-litter partially covered it next to the banks. *Chromidotilapia cavalliensis* was said to be uncommon at this site, but the local people indicated that its distribution was somewhat farther to the north, closer to the source area, i. e. there were it is still shallower and narrower.

Despite many helping hands catching fishes

The Cavally River near Sahibly

with tackles, hand-nets, and drag-nets, no *Chromidotilapia cavalliensis* were caught.

The major part of the catch belonged to the genera *Hemichromis* and *Tilapia*, and in the water-holes near the banks also *Epiplatys* and Tetras were found. The catching procedure turned out very difficult in the wide river. Furthermore we had been warned that since gold was searched for in this area and sand was therefore washed in the water of this river, the desire of the exotic whites to only catch fishes in this river was quite circumspect to many people. Despite all efforts we remained unsuccessful. Eventually we received support from some women who were prepared to catch fishes for us. A short time later a young man appeared with two fish. They were a pair of *Chromidotilapia cavalliensis*, the elusive desirable fish we had come for. Their condition was not very good and the cloudy eyes and dam-

aged fins caused us to exchange the water immediately fortifying it with two drops FMC to 1,5 l water for disinfection. By the afternoon, after quite some time of waiting, the women returned with several bags full of fishes. Besides *Hemichromis elongatus*, *Hemichromis guttatus*, and larger *Tilapia*, several *Chromidotilapia cavalliensis* were amongst the batch. We were explained that the males would carry the eggs in the mouth and the females would not. At the end of the day we had 9 specimens of *C. cavalliensis*. One male had an extensive flesh wound on the tail. All fish were placed in separate container and treated with disinfectant FMC.

After having distributed the presents and paid the women we proceeded back. All *Chromidotilapia cavalliensis* reached their destination in the home aquaria in excellent health and formed the basis for further observations.

Chromidotilapia cavalliensis ♂

Another species of *Chromidotilapia* which is known to exist in various colour-varieties was described in 1974.
In

◗ *Chromidotilapia finleyi*

TREWAVAS, 1974

Colour-morph "Mungo"

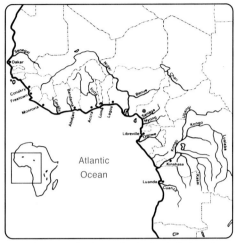

specimens from the river Mungo served as reference material. The males reach almost 11 cm in length whilst the females are fully grown at 9 cm. Typical for all Cichlids from Mungo River-system are the blue to turquoise scale-fields on the flanks and gill-covers. Except for a weakly developed chrome-yellow longitudinal stripe in the dorsal fin, the female has a dark round blotch in the soft rays near the base. Both traits are missing in the male, but his dorsal fin is somewhat longer at the end of soft-ray area of the dorsal fin. Both sexes have

Distribution of *Chromidotilapia finleyi* Colour-morph "Mungo"

pointed ventral fins. Dorsal, caudal, and anal fins are without circumspect pattern.

Chromidotilapia finleyi Colour-morph "Mungo" — top: ♂, bottom ♀

The

natural habitat

of this colour-morph lies between the towns of Kumba and Tombel in western Cameroon. The fish lives in the shallow water zones near the banks and amongst sand-banks in the river Mungo, its northern tributaries, and also the River Menge, the Blackwater River, Kobe River, and Wobe River.

The colour variety inhabiting the northern Mungo River-system between Kumba and Tombel, small shallow watercourses near Ehom in the vicinity of the village Ebonji, is somewhat darker in its over-all colouration. In contrast to most the other areas the ground does not consist of light fine gravel here, but of dark lava-rock.

Here as well as in the Mungo River, the species lives syntopically with the Cichlids *Chromidotilapia guntheri guntheri* and *Chromidotilapia linkei*.

Table 3

Location:	Mungo River, on the road from Kumba to Tombel and Loum, western Cameroon.
Clarity:	feebly murky
Colour:	ochre
pH:	7,6
Total hardness:	3 °dH
Carbonate hardness:	2 °dH
Conductivity:	60 micro-Siemens at 26,5 °C
Nitrite:	0,05 mg/l
Depth:	unknown; near the banks up to appr. 80 cm; width of the river partly 80 to 100 metres
Current:	little
Temperature:	26,5 °C
Date:	12.1.1977
Time:	16.30 hrs

Bridge over the Mungo in western Cameroon

◆ *Chromidotilapia finleyi*

Colour-morph "Moliwe"

This is the third colour variety having an appearance similar to the form "Campo". However, by comparison its colouration is much less intense. The most particular trait of distinction is the black pattern in dorsal fin which partly shows up in the caudal and anal fin as well. The transversal lines which appear during states of excitement are also unique to this variety. In courtship mood these fish display brilliant colours. The belly of the female is of a feeble red colour. The metallic band in the dorsal fin, a feature characteristic for the female in almost all species of *Chromidotilapia*, is hardly or not at all recognizable in this colourmorph. It belongs to the smaller representatives of this species with the males reaching approximately 10 cm and females

Distribution of *Chromidotilapia finleyi* Colour-morph "Moliwe"

9 cm in length. The latter are thus a little smaller.

Chromidotilapia finleyi Colour-morph "Moliwe" ♂

37

The
natural habitat

also lies in western Cameroon. The fish are apparently endemic to a water-course of up to 4 metres in breadth and up to 80 cm in depth only 100 km southwest of the colour-morph "Mungo" crossing the road from Victoria to Kumba between the villages of Moliwe and Mile 4. The water is clean and crystal clear. At some places larger rocks can be found in the stream. The substrate consists of fine quite dark coloured gravel. The adjoining vegetation partly grows deep into the water and provides appropriate hiding-places for the fish.

The following values of the water were obtained by myself in January 1977. The total hardness ranged around 4°dH, the carbonate hardness was established with 3°dH, the pH ranged around 7,6 whilst the measuring of the conductivity showed 130 micro-Siemens at a temperature of 26°C.

Care

Very soft acidic water is required for the colour-morph "Campo" whereas the two other forms can be obviously be kept as easily in moderately hard, alkaline water. These timid fish — all three colourmorphs can be considered as such — should not be kept in aquaria smaller than 130 cm in length. Height and depth should not fall short of 50 cm each. It is recommendable to socialize them only with Cichlid species of the genera *Pelvicachromis* or *Nanochromis*. Cichlids which become somewhat "rough" during courtship periods are inadequate as company. The same tank should however also house fishes of other families, e.g. larger Gouramis or Tetras. It appears to be important that the aquarium is richly vegetated and provides ample hidingplaces, e.g. below pieces of bog-oak or rocks. The water should always be clean and clear which can be achieved by the use of a powerful filter. A constant slight movement of the water is of advantage. Under those circumstances these colour-morphs present themselves without shyness and become interesting objects to study.

Provided with substantial live food, which is readily accepted, young fish grow rapidly and easily become ready for

breeding

which to date unfortunately has rarely been successful. As is the case in all colour-morphs of this species, the form "Moliwe" begins the spawning act with a vigorous shaking of the body in front of the partner with the animals often swimming head-down. In between this behaviour the female cleans a usually obliquely arranged firm surface with her mouth. The light yellow eggs are usually taken in the mouth by the female before the spawning act is completed. It mostly lasts for 45 to 60 minutes. The male prepares a shallow depression in front of the spawning site. Since the eggs often do not stick to the surface, many of them fall into this depression already during spawning and are picked up there by the male. It is mainly the female which carries the eggs in the mouth in the initial stage. Sometimes they are however left back in the pit without direct protection and are thus an easy prey for other fishes.

The behaviour indicates that the transition from a substratum-brooder to a mouthbrooder is not yet completed. There is no doubt that this species belongs to the ovophile mouthbrooders which means the fish take the eggs in the mouth immediately after spawning. As I could observe repeatedly, the spawning procedure is more rigid in the colour-morph "Campo". Here, the female initially takes the eggs in the mouth and passes them on to the male at varying intervals. It may be a couple of hours or even a day until the eggs are forwarded. For this the carrying partner deposits the eggs on the ground from where the other partner picks them up immediately or a few minutes later. This alternating caring is rarely observed and was unknown until a few years ago.

The few successful breedings of this species took place in soft, slightly acidic water. It indicates that not all secrets in the breeding ecology of this species have been revealed yet.

◆ *Chromidotilapia finleyi*
Colour-morph "Campo"

This colour-morph reached the tanks of the aquarists already quite early. I could catch these fish in the forest areas of the Campo Reserve. At this time, they were erroneously considered a possible colour-morph of *Chromidotilapia batesii*. The males attain a length of approximately 12 cm with the females growing a little smaller. A azure-blue narrow longitudinal band in the upper part of the dorsal fin which changes into chrome-yellow towards the base and a fire-red belly of different intensity are the typical feature of the females. In the male, the dorsal fin is bordered with red. There is a blueish glistening blotch bordered with gold ornaments in the hind part of the gill-cover in both sexes. The gill-areas are yellowish golden up to the lips. The remaining parts of the dorsal fin, the anal fin, and the caudal fin lack any distinct pattern and are of a light yellowish golden colour in both sexes. Semi-adult specimens display feebly rosy to frail reddish

Distribution of *Chromidotilapia finleyi* Colour-morph "Campo"

colours and may temporarily have a distinct lateral band on the body. With increasing age the colouration intensifies to a substantial oxide-red.

Chromidotilapia finleyi Colour-morph "Campo" ♂

The

natural habitat

lies in southwestern Cameroon, in the area between the harbour village of Kribi and the frontier town of Campo. It is the region of the river Lobe and the extensive forests of the Campo Reserve. Numerous small water-courses near the coast between the towns of Batanga and Campo mouth into the Atlantic after only a few hundred meters. Measurements taken in this area revealed very soft, definitely acidic water.

Hydrogen-ion concentrations around 4,8 were established. This value is rarely reached in Africa and is in my opinion of outstanding importance for the breeding of these fish in the aquarium at home. The water-courses are running and always carry clear clean water. One might even almost considered it distilled water which frequently mouthes into the ocean in an opaque brown colour. Temperatures range around 25 °C constantly, and occasionally one comes across large assemblies of aquatic plants consisting of *Anubias* and *Bolbitis* which partially grow emerging or submerged on large boulders and are washed over or around by the

Table 4

Location:	Small stream near Eboundja, a village between the towns of Kribi and Campo; southwestern Cameroon.
Clarity:	very clear
Colour:	brownish
pH:	4,8
Total hardness:	below 1 °dH
Carbonate hardness:	below 1 °dH
Conductivity:	10 micro-Siemens at 24 °C
Nitrite:	0,00 mg/l
Depth:	up to 40 cm
Current:	moderate to strong
Temperature:	24,5 °C
Date:	28.3.1973
Time:	11.00 hrs

water. Large colonies of *Crinum natans*, the Water-lily, and especially of *Nymphaea lotus*, the Green Tigerlotus, are found everywhere.

Chromidotilapia finleyi Colour-morph "Campo" ♀

The species of this genus most familiar to aquarists is

▶ *Chromidotilapia guntheri guntheri*
(SAUVAGE, 1882)

An unjustified emendation of the species name is quite frequently observed in many papers which should be adjusted. The French scientist SAUVAGE described this species in 1882 as *Hemichromis guntheri* and used only the letter "u" when naming the fish. In the catalogue of the freshwater fishes of Africa of the English ichthyologist G.A. BOULENGER in which summaries of many African species were published, the mis-spelling of "ue", i.e. *guentheri*, appeared for the first time. It is most likely that this is the source of all subsequent mis-spellings. It is however wrong and the name is to be spelled correctly as *Chromidotilapia guntheri*.

Distribution of *Chromidotilapia guntheri guntheri*

Since this species has a fairly wide distribution, various sizes have become known.

Chromidotilapia guntheri guntheri ♂

41

Generally, one finds indications that this species would grow up to 20 cm or even larger. I would consider these statements doubtful, but it is a fact that this fish reaches approximately 18 cm in the larger rivers of Togo, Ghana, and Ivory Coast. These are imposing males whilst the females hardly exceed 15 cm in length. I could purchase specimens of this size from an African in Ghana which were just hooked from the Birim River near Kibi. In contrast, specimens from Nigeria remain considerably smaller and hardly grow over 12 cm. This length should form the basis for a proper sizing of the aquarium.

Distinguishing the sexes is easy. The females display a distinct chrome-yellow longitudinal band in the dorsal fin and a wine-red belly. Specimens inhabiting the area of the river Mungo not only grow smaller but also differ in their colouration considerably. They reach approximately 10 cm in length. *Chromidotilapia g. guntheri* is often compared with *C.*

kingsleyae or even considered identical. There are however clear points of distinction showing up in colouration and pattern. Directly compared, *C. g. guntheri* has a shorter snout than *C. kingsleyae,* a different number of gill-rakers, generally more spines and fewer soft rays in the dorsal fin, fewer soft rays in the anal fin, a slightly higher number of scales along the lateral line, a slightly lower body-height, small black spots in the area of the spines of the dorsal fin only in the female, and none of the only very feeble blotches or spots in the anal, caudal, or in the area of the soft rays of the dorsal fin.

The

natural habitat

is mostly collectively indicated as West Africa which means Sierra Leone to Congo. In my opinion this distribution range should be restricted and split up. The country of Ivory

Chromidotilapia guntheri guntheri ♀

Coast certainly is the westernmost range limit and, being mainly a region of savannas, produces larger sized specimens up to the west of Nigeria. In this area, the fish usually lives in smaller and larger rivers, lakes, and flooded areas. They may be considered savannah populations. The region between eastern Nigeria and southwestern Cameroon, where in my opinion the southernmost border of the range is reached, is inhabited by smaller growing populations of this species. Here, they live in mostly small to tiny rivers and also often in narrow watercourses which are situated almost exclusively in rainforest areas. This suggests that this species grows larger in its savannah populations usually living in harder and in part slightly alkaline waters of the fields and savannas. It is even occasionally caught in areas with brackish water.

The rainforest populations on the other hand almost exclusively inhabit the soft, often very acidic, clear waters poor in salts. It was however surprising that even specimens of the rainforest areas could be bred in moderately hard alkaline water in captivity without problems and produced numerous descendants. This shows how adaptable the species is.

Care

Keeping this beautiful and mostly timid species generally presents no problems. Aquaria with minimum sizes of 130 cm in length, 50 cm in depth, and 40 cm in height are appropriate. Fine gravel with a grain-size of up to 3 mm should be used. Caves are created by rocks and pieces of bog-oak roots which are placed directly onto the bottom plate before the gravel is filled in. A rich vegetation is recommendable. In front of the built-in caves unobstructed spaces should be taken into consideration since the fish love to dig out "the cave of their choice" by themselves and usually bury smaller plants under the rubble. The

Chromidotilapia guntheri guntheri ♀ from the Mungo River

water should always be clear and clean and a good filter should produce a slight current at all times. One should refrain from planting lawn-like growing foreground plants since *C. g. guntheri* often chews through the substrate in search of food and consequently uproots the plants. Nutrious live food such as mosquito larvae, enchytraeas, larvae of ephemera, small earthworms (rarely daphnia) should form the diet. Flake food is eagerly accepted when offered occasionally. Young specimens grow rapidly under these conditions.

When a pair has bonded, all other specimens of the same species should be removed from the tank unless it is a really large aquarium. This is to prevent territorial disputes and subsequent injuries. Fishes of other families or species of the genera *Nanochromis* or *Pelvicachromis* should on the other hand always be kept alongside since this causes the parents to be especially wary and alert.

Stages in the Reproduction of Chromidotilapia guntheri guntheri

The female lays the eggs onto a sound surface.

Immediately after spawning the eggs are fertilized by the male.

The male alone takes the eggs into the mouth for incubation.

The juveniles are cared for by both parental specimens.

The

reproduction

is initiated by the females shaking the body in front of the partner. The females gleam in very intense colours at this time. *Chromidotilapia guntheri guntheri* belongs to the ovophile mouthbrooders. The spawn is deposited on a stone or another solid surface and thereafter is taken into the mouth by the male. It is only the male which incubates the offspring. Thereafter, the female joins in to protect and guide the young fish. Both parents take the fry in their mouths in cases of danger or order the small ones to stay motionless by jerkily erecting the dorsal fin or spreading the ventral fins. Despite agreeing statements in various publications about the incubation period of the young fish

lasting 10 to 14 days, I could observe the male spitting out the fry already after 6 days at a water-temperature of 26 °C. They were picked up by the female for a short while and could already swim fairly well. Eight days after the spawning, all the young fish were swimming freely and first food was accepted. The off-spring usually numbers more than 100. Immediately after being released from the mouth they accept the nauplii of the Brine Shrimp *Artemia salina*. The parents devotedly lead and care for their descendants for several weeks. Provided sufficient quantities of food and a high quality of water, the young fish grow rapidly and may measure about 20 mm already two weeks after the first swimming attempts.

Are fishes supposed to be here? A water-course in Nigeria.

◗ *Chromidotilapia g. guntheri*

Colour-morph "Bosumtwi"

Only in March 1988, after a visit to the Lake Bosumtwi, the first ten specimens were brought back to Europe. Before this, they could not be observed in captivity. It was interesting to note that they were slightly different from *C. g. guntheri*. Visually, the slightly more pointed head and the partially yellow coloured iris of the eye are feeble points of distinction. These fish are also mouthbreeders, with the male incubating the offspring and both parents jointly caring for the juveniles.

The

natural habitat

of Lake Bosumtwi lies in the West African state of Ghana. It is almost circular-shaped, and measures maximally 8 km in diameter corresponding to the 6 km of the narrowest distance. It greatest depth is 76 m in the centre. The lake is surrounded by an almost 100 metres high crater wall isolating it from the

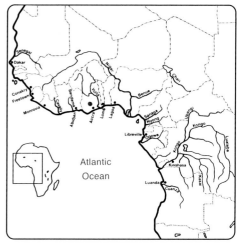

Distribution of *Chromidotilapia g. guntheri* Colour-morph "Bosumtwi"

surrounding. During the rainy season, it is fed by 37 small and larger tributaries. Approximately 20 small villages and settlements are situated on its shores. The only access is provided by an old bumpy track over the crater

Chromidotilapia g. guntheri Colour-morph "Bosumtwi" ♀

wall. For religious reasons it is forbidden to step into the lake, and even the local fishermen may enter it only on so-called swimming stakes. The only two available wooden boats may be used only in cases of emergency for the medical treatment of the residents and with the permission of the Ashanti King of Kumasi.

This lake originates from a meteorite striking the area in prehistoric times. The isolation has caused a separated evolution of the underwater fauna so that one may talk about several endemic species of fishes in this case. Will the Lake Bosumtwi in the Kingdom of Kumasi still keep its mystically enveloped secrets for some time? Some travel notes:

The drive of approximately one hour and 15 minutes to the lake from Kumasi via Jachi, over an old tar-road and partly on a dirt-track with numerous potholes, is uphill all the way. It leads through a hilly rainforest area with cultivations of oil-palms, bananas, and pineapple in loose succession. At approximately 10 o'clock we reached the edge of the crater presenting the Lake Bosumtwi as an impressive water-surface. A spectacular view. The lake almost resembles a shining, almost circular, spread-out mirror which is surrounded by an equally high mountain-chain whose rock is held together by brush and trees. Silence encompassed us. No cars near or far. Because of the height on the crater edge there was some wind. It was pleasantly crisp with the air-temperatures being only 28 to 30 °C. Of the many small villages, only Abonu can be reached over a 3,5 km long serpentine track. It is a small fishing village on the northern shore of the lake. It happens very seldom that a car comes here. Occasionally a mini-bus stops here which is usually the only connection to the world outside.

The lake lay very silently in the sun. A weak wind caused a little movement to the water.

Float-beam on the shore of Lake Bosumtwi

This valley basin reacts like a greenhouse to the exposure to the sun. On the shore, the temperatures ranged between 35 and 36 °C. The heating of the water is considerable under the strong sun. The few streams which supply water to the lake during the rainy season run down the mountain slopes and the rain water carries a lot of sediment. This in conjunction with the high rate of evaporation of the water of the lake lead to a high mineral content. The total hardness of 4 °dH was a surprise. The carbonate hardness on the other hand ranged around 27 °dH and the conductivity varied between 880 and 907 micro-Siemens at water-temperatures between 30,4 and 31,2 °C. Unusually high however was the pH of 9,2. The water was clear to slightly murky.

Following the statements of the local fishermen of Abonu there are six different-looking fishes in the Lake Bosumtwi. As far as is known all of them are Cichlids, made up of four species of *Tilapia* and one species of *Hemichromis* and *Chromidotilapia* each. All adult specimens we saw reached only 12 cm in length. Due to the isolated evolution one may presume that all species living here reach a smaller adult size in comparison with similar or related river populations of the surrounding area.

The fishermen told us that *Chromidotilapia g. guntheri*, colour-morph "Bosumtwi", does not live in the lake normally, but would inhabit the very short rivers supplying the lake. During the rainy seasons the large amounts of water would flush the fish into the lake where they however could not survive for a long time and could not be recorded permanently. They would inhabit the short tributaries also in the dry season, but not be present in the lake then. The water values there may possibly be more adequate for this species. No measurements were however taken.

The *Chromidotilapia*-form of this habitat apparently grows smaller in comparison and is fully grown at 10 to 12 cm. *Chromidotilapia guntheri* from the Birim River on the other hand could be recorded with total-lengths of 16 to 18 cm. All species found in the Lake Bosumtwi as adults showed little differentiation in comparison with similar or related species inhabiting the rivers outside the Lake Bosumtwi crater.

Chromidotilapia g. guntheri Colour-morph "Bosumtwi" ♂

A subspecies of *C. guntheri* which apparently has not yet been imported alive is

▶ *Chromidotilapia guntheri loennbergi*
(TREWAVAS, 1962)

It lives endemically in the crater-lake Barombi-ba-Kotto and its mouth into the Nganjoke River, in the Mungo River-system near Kumba, and the nearby small crater-lake Mboandong in western Cameroon. This form usually has only 14 scales around the caudal peduncle whereas usually all *Chromidotilapia*-species have 16 scales. This form however also differs distinctly from the other *Chromidotilapia-species* and only *C. linkei* partly shows some parallels. In order to give an idea about this form, a summary of the original description shall be given here.

The ground-colour is lighter than in all other fishes from the Mungo area and may be compared with yellow sand. Laterally, from the hind parts of the head to the tail region, the colour changes from a watery green to yellow darkening somewhat towards the lower belly. A black gill-blotch is present. The lower parts

Distribution of *Chromidotilapia guntheri loennbergi*

of the head are brilliant white or coloured rosy. The chest portion is white in the female and rosy in the male with the belly and lower lateral parts often having a feeble rosy colour. Dorsal and caudal fins have a yellow ground-colour. The edge of the soft part of the dorsal fin and the upper tip of the caudal fin is black with a white border on the inner side in large specimens. The spinous part of the dorsal fin bears numerous small black spots in one to three irregular arrangements which may extend up into the soft-ray portions. These spots may be underlaid by a broad silverish bronze coloured band surrounded by grey and black in mature females. The males lack these spots and the silverish bronze band. Instead, they may have feeble light grey blotches on a yellow ground in the soft part of the dorsal fin. The caudal fin usually lacks spots in the male but not in the female. The anal fin is grey, sometimes with a rosy tinge. The pectoral fins are light yellow.

After this colour description it is to be hoped that these probably very beautiful fish will find the way into our aquaria one day.

A large specimen of a species of *Anubias*

49

◆ *Chromidotilapia kingsleyae*
BOULENGER, 1898

This species has been known to aquarists for a long time and many publications in journals and magazines deal with the husbandry and breeding of this largest representative of the genus *Chromidotilapia*. However, I dare to doubt that it is frequently kept in the tanks of the aquarists. Unfortunately this species is often confused with *C. g. guntheri*. The points of distinction are not always clear in subadult specimens, but very clear in adult animals. Besides the morphological differences there are other particularities in colouration and pattern. Compared with *C. guntheri guntheri, C. kingsleyae* has a longer snout, a different number of gill-rakers, generally fewer spines in the dorsal fin and more soft rays in the anal fin, usually a lower number of scales along the lateral line, the body is somewhat higher, small black spots in the spinous part of the dorsal fin are absent in the female, but both sexes have

Distribution of *Chromidotilapia kingsleyae*

small black spots in the posterior part of the soft portion of the dorsal fin, between the posterior rays of the anal fin, as well as on the entire caudal fin. Fully grown males display a

Juvenile specimen of *Chromidotilapia kingsleyae*

light silverish colouration with a longitudinal row of blueish black spots. Large specimens are said to reach 20 cm in length. In this species the females also stay slightly smaller. Under captive circumstances these fish attain usually only 12 to 14 cm in length.

The

natural habitat

was indicated by G.A. BOULENGER on the basis of collected specimens as being Niger, i.e. the River Niger in Nigeria, up to Congo. Only two specimens originated from the Opobo River in the Niger Delta. The other 24 reference specimens came from Gabon and the Shiloango region, the frontier river between Zaïre and the small enclave of Angola. It appears that these are the only specimens which have ever been caught so far north. It is likely that this was based on an error already then and in fact referred to *C. g. guntheri*. Single specimens of *C. kingsleyae* were repeatedly caught in southeastern Cameroon

near the frontier with Gabon and these are thus the northernmost representatives of this species.

Care

Large tanks are required für these fish. Only then they display themselves in all their beauty and are very interesting objects to study. According to the numerous existing reports, they belong to the ovophile mouthbreeders. Their breeding and husbandry conditions are comparable with the species *C. g. guntheri*. Well and densely planted aquaria should offer many hiding places between rock constructions and bog-oak arrangements. The water should be clear and clean, and be of soft and slight acidic quality. A constant slight current of the water should not be overlooked. Temperatures around 26 °C are ideal. Potent live food, such as mosquito larvae, earthworms, enchytraeas in addition to flake food, should be obligatory because feeding a variety of foods is very important for this Cichlids.

A branch of the small Ogba River in Nigeria

◗ *Chromidotilapia linkei*
STAECK, 1980

This very pretty coloured Cichlid became known only relatively recently. The males reach approximately 10 cm in length with the females staying only slightly smaller. The sexual dichromatism is very clear. The females bear a metallic chrome-yellow "iridescent" longitudinal band in the spinous part of the dorsal fin. A distinct feature of this species is the brilliant light blue iris of the eye as well as a large dark gill-blotch which is surrounded by a light blue area especially in the female. This colour also shows on the lateral scales of the body — an appearance also known from other species of *Chromidotilapia* from the Mungo area. The lower lip is coloured pearl-white.

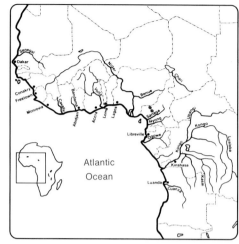

Distribution of *Chromidotilapia linkei*

Chromidotilapia linkei — top: ♀, bottom: ♂

This species is timid and can be kept together with other fishes very well. It is territorial throughout the year.

The

natural habitat

lies in the region east of the 4010 m high Mount Cameroon in the Mungo River and its tributaries in western Cameroon.

During an expedition in the year 1977, I was able to catch this species in the shallow water zones of the Mungo River. These were approximately 10 metres broad strips along the banks which were partially up to 50 cm deep and connected to the main stream only by narrow gullies during the seasons of low-water. The substrate consisted of fine gravel up to a grain size of 3 mm and was bare of plants. Branchwork and driftwood alongside with bank-vegetation growing into the water provided hiding and staying places for the fishes living here. Catches made also contained two other species of *Chromidotilapia* which inhabited the Mungo River near the bridge of the road from Kumba to Tombel. These were *Chromidotilapia finleyi* and *C. g. guntheri* which were different from specimens of other areas in size and colouration. In the nearby areas around the village Ehom I found the same composition of species of this genus. Whilst the substrate of the Mungo River consisted of a light coloured fine gravel, the areas around Ehom are constituted by the brownish laterite soil of the Mount Cameroon. A lake-like water-body behind the village, situated in the banana plantation, is used as freshwater supply. The depth of the water is approximately 1 metre, and large lava rocks form sites of rubble at some places. Green species of *Nymphaea* and dense groups of a tropical horn-weed as well as knifesharp edged assemblies of reed form the submerse vegetation. The water is very clear and somewhat torrential even here. Whilst the water-values in the bank zones of the Mungo River were established as 3 °dH total and 2 °dH carbonate hardness at a pH of 7,6 and a conductivity of 60 micro-Siemens at 26,5 °C, those near Ehom were 170 micro-Siemens at 25 °C with a

hydrogen-ion concentration of 7,8. These values are of interest for a husbandry in tanks with a moderately hard water from the tap. At both places *Chromidotilapia* species were numerous. Upon catching *Chromidotilapia linkei* in the Mungo River I already noted that this species is a mouthbreeder; upon being transferred into a plastic-bag, an approximately 9 cm long female spat out some 20 swimming juveniles which were unfortunately immediately eaten by other fishes.

Care

The following recommendations can be made for the husbandry of this species. The chosen aquarium should not be too small. Sizes of 130 cm in length, 50 cm in depth, and a height of 40 cm are adequate, but larger tanks are preferrable. Fine and slightly dark gravel with a grain-size of up to 3 mm should be used as substrate. Rounded stones which are placed directly onto the bottom plate serve for building caves. Pieces of bog-oak also provide appreciated sites to stay. The aquarium for *Chromidotilapia linkei* should be richly vegetated. Since this species is tolerant towards plants, one may decorate and plant as one desires. If the West African natural biotope is to be emulated, aquatic plants of the genus *Anubias* are the choice for the background. Especially *Anubias gilletii, A. barteri, A. afzelii,* and *A. heterophylla* are suitable for this purpose. They should be arranged in groups. In between, or even better in gaps in front of them, Red and Green Tigerlotus, *Nymphaea lotus,* may be planted. The sides and decoration items such as rocks and wood, can be furnished with the Congo-fern *Bolbitis heudelotii,* interspersed with the small-growing lanceolate-leaved *Anubias nana* which may also serve to plant the foreground. One may also attach these plants which reach only 8 cm in height by glasspins to the previously installed styrofoam rear- and side-walls. The *Anubias* and *Bolbitis* plants must be buried with the roots only whereas the rhizomes are to stick out free. The water should not be too hard and range in the neutral zone of the hydrogen-ionconcentration. A pH around 7 is recommendable. The water should be clean and rich in oxygen at all times. A slight current initiated by a powerful pump, is of importance. The watertemperatures should range around 25 °C and the illumination be switched on for 12 hours. The length of the day equals the length of the night in the tropics. Plants require light for growth and healthy plants mean healthy water. Healthy water in turn means healthy fish.

Notwithstanding this, a quarter to a third of the water should be exchanged with fresh water every week or two in a "settled" aquarium which can be considered as such after four to six weeks. Nutrious live food, including large daphnia from time to time, should form the diet. Flake-food is readily accepted as well. Provided these conditions, young *Chromidotilapia linkei* grow up to large healthy fish within one year. For a stress-free environment one should remove all specimens from the tank except for one harmonizing pair. Nevertheless it is recommendable to keep small specimens of other Cichlid genera and larger Gouramis or Tetras with them.

The

breeding

with a harmonizing pair is easy and has been successful several times. The species belongs to the ovophile mouthbrooders which means that the eggs are taken in the mouth right after spawning. Both parents participate in incubating the offspring. They spawn on a solid surface and both partners defend the spawning territory. If several other fishes are around and threaten the spawn, the male takes the eggs into his mouth already during the spawning act. If there is no danger, he waits until the female has completed spawning and then puts all of them together. In the first described case the male places the eggs back at the spawning site or in the immediate vicinity towards the end of the spawning act from where the female picks them up at once. The eggs are of oval shape, measure 2,4 mm in length, and have a yellow colour. Thereafter the female retreats to a calm and covered area in the aquarium and is protected by the male in cases of danger. Also

during this time there is much harmony amongst the sexes which may last for many years even if one partner is removed for a while and then re-introduced. The female carries the eggs, and the subsequent larvae, for approximately seven days. They are then passed on to the male which carries them for another week. At a water-temperature of 25 °C during the incubation period, the juveniles are released from the mouth of the male as swimming young fish on the 13th or 14th day. They immediately start feeding on newly hatched nauplii of *Artemia salina*. The parents care for their fry another three or four weeks until the young fish have grown up to some 2 cm. They only number 20 to 25.

These breedings occurred in water provided by the Berlin (Germany) waterworks having a total hardness of 15 °dH, a carbonate hardness of 12 °dH, a pH of 7,9, and a conductivity of 475 micro-Siemens. It would be of interest to find out whether water-values corresponding with the natural habitats would result in a higher number of juveniles since approximately 150 eggs are originally taken in the mouth for incubation. My results numbered 23 young fish on average. Experimentally removing the male from the tank with a fine net and transferring it into another aquarium on the tenth day of breeding resulted in the fish spitting out a couple of larvae which were however taken back again. The larvae measured 7 mm at this point of time and their eyes were already fully developed. Mouth, gills, anus, and heart were easily recognizable.

The sequence of spawning and development of the fry summarized here is based on own observations. I however deem it possible that variation may occur.

Only known from some preserved collection material,

◆ *Chromidotilapia schoutedeni*
(POLL & THYS, 1967)

lives in the east of Congo. The reference material originated from Shabunda, Yangambi near Lubilu, and Lomboma, Stanleyville, in the area of the Lualaba. The first glance reveals a fish unlike all other *Chromidotilapia*, and its appearance lets one rather presume it would be a species of *Pelvicachromis* or *Parananochromis*. But the drawing contained in the revision by THYS falsifies the ratio of the total lengths of the sexes. As far as is known so far, the males attain approximately 9 cm in length with the females growing a few millimetres smaller. Also in this species, the female has a shiny metallic band in the dorsal fin which is missing in the male. The species appears more slender and less high and has 16 scales around the caudal peduncle. One point of distinction from other species of the genus *Chromidotila-*

Distribution of *Chromidotilapia schoutedeni*

pia is especially noteworthy. Unlike the males, the females of this species have rounded ventral fins, a feature which is characteristic to species of the genus *Pelvicachromis*.

The approximately 10 cm long fish illustrated here is probably a male specimen of *Chromidotilapia schoutedeni*. It was imported as a single specimen from the region of Lualaba, north of Kindu.

Another species which was only recently imported alive is as well little known.

◆ Chromidotilapia sp. "Shiloango"

was not only found in the area of the Shiloango River, the frontier river between Zaïre and the small northern enclave of Angola near the coast of the Atlantic, but is also said to live in Niari, Kouilou and Nyanga in Gabon and Congo respectively. The maximum total length so far recorded ranges up to 12,5 cm.

For this fish, Roland NUMRICH noted "we could record these animals in some savannah tributaries of the Nyanga as well as in some independent water-courses of the coastal rain-forests. *C. kingsleyae* is found in the main

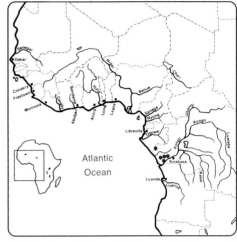

Distribution of *Chromidotilapia* sp. "Shiloango"

Chromidotilapia sp. "Shiloango"

drainage system of this area, between the Du-Chaillu-Massif and the coast, the Ngounie River. The biotopes of this 10 to 12 cm long species are smaller rivers and torrential streams whose ground consists of a light and very fine sand. Plants are rather rare, occasionally groups of water-lillies, *Crinum* sp., are found whose up to two metres long leaves float in the water and provide cover for many Top Minnows of the genus *Plataplochilus*. The *Chromidotilapia* usually stay together in loose schools of 10 to 12 specimens in the vicinity of underwashed bank areas where they find excellent cover amongst the mounds of leaf-litter, lush branchwork, or roots of the bank's vegetation.

Catching these fish therefore turned out to be a difficult task thus we ended up with having local people catching them with fishing rods.

Despite the proximity to the coast, the water of these rivers is very soft and acidic (pH 5,2 to 5,8, conductivity 20 to 35 micro-Siemens) and quite cool at 22 °C. The major source of food for these *Chromidotilapia* is probably a 5 to 10 mm long black and red coloured shrimp which occurs here in ample numbers. It vaguely resembles the Asiatic Zebra-shrimps of the genus *Caridina*, but has black spots instead of bands."

According to observations made so far, these fish attain a length of approximately 13 cm. Females grow up to a mere 9 to 10 cm and are thus considerably smaller. They are ovophile mouthbreeders with both parents alternatingly incubating the offspring and devotedly caring for their fry. So far, soft, slightly acidic to neutral water has been proven adequate for the husbandry and breeding of this species. Their reproduction in captivity has been successful on numerous occasions. Only the process of assembling harmonizing partners turned out to be difficult.

Compared with other species of the genus *Chromidotilapia* these fish are not as high and thus appear more slender.

Dark lava-like rocks cover the ground of the water-courses in the Kumasi area in Ghana.

◆ Chromidotilapia sp. "Mondemba"

This species was discovered only in 1980 and is yet to be described scientifically. It occurs in the frontier area between Nigeria and Cameroon, in the region of the Korup River (Akpa-Yafe) which, over a long distance, forms the border between both these countries. It may possibly be the smallest *Chromidotilapia*-species which apparently does not exceed an adult-size of 8 cm.

The fish was imported alive by M. FREIER only in 1989 for the first time, and the few specimens originated from the Ndian River in the region of the Rumpi Hills near the town Mondemba (Mundemba) in western Cameroon. This species seems to be endemic to this area.

According to what has been experienced so far, the husbandry of this fish is not connected to specific problems. Its breeding however appears to be a problem with regard to the assembly of a well harmonizing pair. This in

Distribution of *Chromidotilapia* sp. "Mondemba"

turn means that the partners pick each other out of a large group of specimens.

Chromidotilapia sp. "Mondemba" ♀

In the year 1988 KLAUS LANDSBERG and JÖRG WUNDERLICH managed to discover an entirely new Cichlid during an excursion through Gabon and even to bring it home to Europe alive. This is another representative of the genus *Chromidotilapia* which shall be referred to here as

▶ *Chromidotilapia* sp. "Atogafina".

The males of this scientifically undescribed species grow up to 15 cm under aquarium conditions whilst the females attain a total lengths of around 12 cm only. Since this species is thus a somewhat larger growing fish, aquaria of appropriate sizes are required. The representatives of this genus have the most interesting breeding behaviour of all West African Cichlids. As far as has been established so far, all of them are mouthbreeders which have developed different strategies. Often it is only the male which incubates the fry in the mouth, but in some species both partners alternatingly care

Distribution of *Chromidotilapia* sp. "Atogafina"

for the offspring. In this particular species however, it appears to be the female which carries and incubates the eggs in the mouth. Another very interesting point of distinction is found in the outer appearance.

Chromidotilapia sp. "Atogafina" ♂

This very pretty and colourful new species is the only one where the male has a lyreate caudal fin. No other form of *Chromidotilapia* has a caudal fin of this shape in either male or female. The fish were caught in the area of Atogafina, approximately 50 km northwest of Libreville, the capital of Gabon. The locality is a small, approximately 50 cm deep, clear water-course with a slight current in a some-what hilly environment. By having 39 micro-Siemens the water was very poor in minerals and with a pH of 5,8 quite acidic. The water-temperature was approximately 23 °C (pers. comm.). These pretty, timid, and somewhat shy fish should be referred to as *Chromidoti-lapia* sp. "Atogafina" until a formal scientific description is published.

Depending on the mood and environment, these animals display light brownish pastel colours on the body or a light beige to sand-colour, or even light yellow shades, with irregularly arranged large black blotches. They however always show a dark red edge on the fins which is partially conspicuously bordered with light blue making the species especially attractive.

Jewel fishes which would form the other complex. Due to limited space I want to abandon on listing the species alphabetically here, but quote the species-groups as suggested by LOISELLE:

Species-group 1:
Hemichromis fasciatus
Hemichromis elongatus
Hemichromis frempongi

Species-group 2:
Hemichromis bimaculatus
Hemichromis cristatus
Hemichromis paynei

Species-group 3:
Hemichromis guttatus
Hemichromis stellifer
Hemichromis cerasogaster

Species-group 4:
Hemichromis letourneauxi
Hemichromis lifalili

After the revision by the American ichthyologist LOISELLE in 1979, the

Genus Hemichromis

contains eleven species, five of which were described as new or resurrected from the synonymy into valid species rank in the work "A Revision of the Genus *Hemichromis* PETERS 1858".

LOISELLE divided the eleven species into four species-groups which in my opinion should however be split into separate complexes or rather into two separate genera. This is the *H. fasciatus*-group which should be assigned to one complex due to size, appear-ance, and behaviour and the so-called

Even very shallow water-courses are habitats for larger Cichlids, as is the case in these *Hemichromis elongatus*.

◆ *Hemichromis bimaculatus*

GILL, 1862

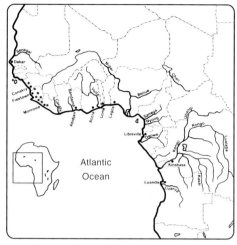

With this species, we begin with the portraying of the so-called "Jewel fishes" in alphabetical order. Nine species are recognized so far, but it is unknown whether all of them have been kept in captivity as well. All forms can only be kept together with smaller Cichlid species under certain conditions. They are not really aggressive, but, as is the case in all Cichlids, one should expect some aggression during the reproductive seasons. So far, all Jewel fishes were referred to as *H. bimaculatus*. It was not before the erroneous description of *H. fugax* that it turned out that the real *H. bimaculatus* is distinctly different from all other forms in its appearance and body-shape. As the name

Distribution of *Hemichromis bimaculatus*

Hemichromis bimaculatus ♂

suggests, this fish displays two blotches (bi = two, macula = blotch) on the lateral sides of the body. This does however not include the dark blotch on the gill-cover in any species. The species appears relatively elongate with the long snout causing a different shape. The ground colour is light beige. The shock-pattern consists of several narrow transversal lines. Only when the fish feels especially well or during the courtship- and spawning-season the colouration is changed and the major parts of the body become brownish beige to oxide-red. The lips, the chin up to the chest, as well as the belly are fire-red with the dark gill-blotch bordered by a golden colour. Only the males display several light, usually golden yellow spots on the gills. They reach 11 cm in length. The females grow a little smaller and have a wider girth. So far, this species has been kept in captivity only on rare occasions.

The

natural habitat

lies in Sierra Leone, Guinea, and Liberia. The fish usually inhabit small to minute watercourses with clear, soft, acidic water. Self-caught specimens of this species always were sound brownish beige to oxide-red and displayed a lot of red colour even in the fins. At some places they share their biotope with *H. paynei*. The specimens illustrated here were caught in the forest areas of Kasewe.

Hemichromis bimaculatus ♀

◆ *Hemichromis cerasogaster*
(BOULENGER, 1889)

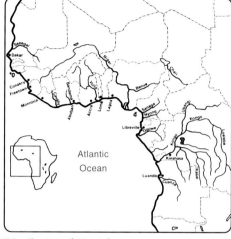

Distribution of *Hemichromis cerasogaster*

has probably not yet been imported. The species was previously assigned to the genus *Pelmatochromis*. It shows a lot of violet and orange-red colour. Depending on the mood, one may recognize approximately ten light vertical bands which extend from the dark upper area downward and end in the light section. Between approximately the 10th and 14th spine of the dorsal fin there is a dark zone from the outer edge to the base. Except for a dark edged blotch on the gill-cover there are no other lateral blotches present which otherwise are so typical for the genus. These fish attain approximately 10 cm in length.

The

natural habitat

is the Lake Maji-ndombe, previously known as Lake Leopold II, in Zaïre.

Establishing water-qualities is one of the most important tasks during an expedition.

♦ Hemichromis cristatus

LOISELLE, 1979

is a colourful species and shows the conspicuous rows of usually light blue spots typical for the members of the genus all over the body. A characteristic feature to distinguish this species from its allies may be the light periphery around the dark central blotch. Furthermore it lacks the distinct gill-blotch. The sexes can be accurately distinguished only during the spawning season by the female having a slightly wider girth at this time. Mostly the females grow a little smaller than the males. The species attains a length of approximately 9 cm and is timid.

The

natural habitat

lies in small streams in the countries of Guinea, Ghana, and especially Nigeria which usually contain soft, acidic water. The reference specimen originated from the Ogba River in Nigeria. It is a small river crossing Benin City in the region of the Niger Delta. The water-

Distribution of *Hemichromis cristatus*

temperature ranged around 27,5 °C in March. Total and carbonate hardness were below 1 °dH with a pH of approximately 4,8. The conductivity was measured with 5 micro-Siemens at 27,5 °C. In 1978, this species was abundant in the Ogba River where the specimens illustrated here were caught.

Hemichromis cristatus

◆ *Hemichromis elongatus*
(GUICHENOT, 1861)

This species differs from the afore mentioned in its colouration and behaviour. Furthermore, it grows somewhat larger and is fully grown at some 15 cm. A major point of distinction is the greenish silver ground-colour which is especially conspicuous between the chin and the tail in the lower half of the body in adult specimens. In contrast to *H. fasciatus* and *H. frempongi*, its fins are coloured smoky grey. The five roundish dark blotches on the flanks may transform into dark vertical stripes. During courtship and spawning the males assume a sound dark green colouration which may even sometimes temporarily turn into black. Despite all this beauty one should never forget that *Hemichromis elongatus* is a predatory fish.

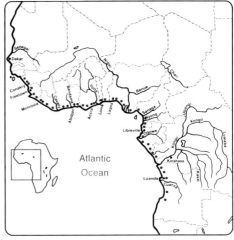

Distribution of *Hemichromis elongatus*

The

natural habitat

extends from Guinea through Sierra Leone throughout the entire West African coast-line up to Congo, Zaïre, and Angola. This fish has even been caught in Zambia.

Hemichromis elongatus

◗ *Hemichromis fasciatus*

PETERS, 1858

is another predatory fish which is aggressive and antisocial — a widely accepted judgement about this species. Notwithstanding *H. fasciatus* is an interesting object to keep and to study. Rarely one comes across a fish where speed, eagerness for food, and intelligence — as far as one can consider fishes as intelligent — are so obvious. Self-confidence and the will to survive are also very prominent. On the other hand, these attributes represent a threat to its own species, respectively in the absence of natural predators result in a control of the population density. All three species of this group are lone-wolves. Until maturity is reached everything that can be fought and overwhelmed is claimed. Even almost same-sized siblings may become victims of this eagerness. If two mature specimens meet and

Distribution of *Hemichromis fasciatus*

are willing to spawn, they form a pair living in harmony. They enter into a lasting partnership,

Hemichromis fasciatus

occupy a territory, and may only tolerate small co-occupants of other species. In the wild, the territory may cover an area of 18 to 25 square metres. Under the conditions of captive husbandry, a tank has therefore to be of really large size if other fishes, of different species and genera, are to have a chance to survive. Husbandry without risks is thus only possible in the case of individual specimens or a harmonizing pair during spawning-season. To see if this is the case it should however be tested beforehand by introducing the potential partners to each other through a separating plate of glass. If both show willingness to spawn, the glass plate can be removed. Nonetheless one should monitor what happens then.

This species temporarily shows five large roundish dark blotches on the flanks, the first of which is behind the gill-blotch and the last

on the caudal peduncle, sometimes with additional small dark spots around the latter. The ventral side of the body, between the chin and the central part of the tail, i. e. the centre of the anal fin, temporarily assumes an intense red colouration. The back is beige to dark brown with the lateral sides being silverish. Caudal and anal fins are usually transparent. The species attains a length of approximately 18 cm.

The

natural habitat

extends from Senegal through Mali, Sierra Leone, Liberia, Ivory Coast, Upper Volta, the Central African Republic, Chad, Togo, Ghana, the Republic of Benin, up to Cameroon. It consists of large and small rivers and watercourses with a wide variety of water-qualities.

Almost everywhere friendly Africans help with catching fishes.

The third species of the *H. fasciatus*-group is also the one most recently discovered.

▶ *Hemichromis frempongi*
LOISELLE, 1979

differs only slightly from the type species *H. fasciatus*. Its total length may reach approximately 14 cm in adult specimens. The five, usually round blotches on the flanks of the body are smaller and lack the encompassing small spots. This fish temporarily displays an intense red colouration of the entire ventral third of the body from the chin to the caudal peduncle. A faint lateral stripe may connect the first and the last lateral blotch depending on the actual disposition.

Distribution of *Hemichromis frempongi*

The
natural habitat

lies endemically in the Lake Bosumtwi in central Ghana.

Care

Keeping this species may present problems because large tanks are needed. Smaller tanks may be only appropriate if the fish are kept

Hemichromis frempongi ♂ photographed in the field

individually. Even many hiding places do not necessarily help other fishes or females not ready to spawn. The best method is to keep the partner separated by a glass plate until it becomes obvious that both are willing to mate. Even half-grown specimens can be kept together with other fishes only if provided with large quantities of food. The diet should consist of earthworms, fish — of a third to up to the half of the size of these fish — and other nutrious sorts of food.

By reaching a length of approximately 9 cm, they become mature and may start

breeding.

For a well harmonizing pair, a tank of 130 cm in length, 50 cm in depth, and a height of 40 cm should be sufficient. Fine gravel should be used as substrate. Some large flat stones and a few robust plants placed near the background and the side walls make up the decoration. A powerful filter system has to kept the water clean and rich in oxygen. A water-temperature of 25 or 26 °C is suitable. A flat stone serves as spawning-site. It is often cleaned only superficially before the female lays her eggs in intervals of 10 to 20 at a time. They are fertilized immediately. After approximately two hours a total number of 400 to 500 eggs have been laid and the female begins to care for the clutch whilst the male guards the spawning territory. If no smaller company fishes are around, the partners may engage in quarrels. A large tank however offers the opportunity to separate a section of it by a glass plate behind which a few larger fishes should be kept. This enables the male to release its defence aggression. Without the separating glass plate every other fish, no matter if even of larger size, would have to be considered a target. 36 to 48 hours after spawning the hatched larvae are transferred by the female to a well covered pit. Another 90 hours later the fry swims and is usually guided by the female. Large amounts of newly hatched nauplii of the Brine Shrimp serve as first food and should be supplied several times a day. The food-requirements can hardly be satiated. The parental pair is very attentive and devoted to care for the offspring. A strong jerking, i. e. a fast erecting and subsequent flattening of the

dorsal, ventral, and anal fins, is the signal of the parents for a potential danger and causes the juveniles to stay motionless. With increasing age, the young fish become more active and now the male joins in when it comes to collecting the small runaways. These are taken into the mouth and brought back to the school. At dusk, the parents collect the fry in a depression or a cave.

This behaviour can be observed when the juveniles are approximately 2 month old. The offspring would have attained a length of 2 to 3 cm by then and display a dark lateral stripe on the body. The number of juveniles has usually greatly reduced at this stage since the tendency to cannibalism shows up already quite early in this predatory fish. The young will kill their weaker siblings and consume them entirely. This can be even more frequently observed in the wild than in a aquarium, and at a length of 4 cm only very few specimens are usually left. The parental care subsequently ceases and the young fish change the lateral band into a blotch pattern. At this time they are no longer tolerated in the school and the protective effect falls away.

By now being "five-blotched fish", they are recognizable as predatory fish. Spread over a wide area, the individualists have to scavenge for their food alone now. Damaged and underweight caused by constant intraspecific fights, they do not lead a life favouring growth. Five to six centimetres long specimens which I caught occasionally had seriously sunken bellies. Despite the large numbers in the clutches, not many specimens reach full size.

I have observed the closely related *H. elongatus* and *H. fasciatus* in Cameroon, and also in Nigeria, in water-courses of only 10 to 20 cm in depth with a slight current where they were staying motionless for a long time. Since these calm and shallow waters are favoured sites for smaller fishes, they are also the preferred hunting territories for these pretty predatory fish. Notwithstanding their aggression, they are very interesting objects to study.

Hemichromis guttatus
GÜNTHER, 1862

is a larger growing species of approximately 11 cm in total length. It displays a very pretty red colouration which is somewhat lighter in the areas of the gills and on the belly. Whilst the dark gillblotch is very conspicuous, the caudal peduncle lacks a blotch. Only the body bears a dark roundish blotch somewhat above the centre which may be stretched upwards to the back and downwards to the anal region depending on the disposition and may thus form a broad stripe. This stripe is however less intense outside the dark centre. It may sometimes be encircled by a light corona. The body appears to be somewhat higher than in allied species.

Distribution of *Hemichromis guttatus*

The

natural habitat

lies in the countries of Ivory Coast, Ghana, Nigeria, and Cameroon.

Ferries connect most of the banks in the southern West Africa today.

◆ *Hemichromis letourneauxi*
SAUVAGE, 1880

The species of the genus *Hemichromis* are not always true "Red Jewel Fishes". Instead, olive beige or brownish violet colours intermixed with grey shades are found in certain species. *Hemichromis letourneauxi* belongs to this group. It reaches a total length of approximately 15 cm, and thus is a large "Jewel Fish". Regarding its behaviour and husbandry requirements it can be compared with the other species. It is not of outstanding aggression, and keeping it together with other fishes of approximately same size does usually not cause problems — provided the aquarium is large enough. If however a pair becomes ready to spawn, a large territory is claimed. Distinguishing the sexes is as difficult as in all other species of this genus.

The

natural habitat

extends over the northern subtropical region

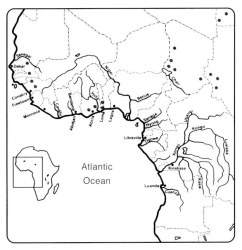

Distribution of *Hemichromis letourneauxi*

of Africa. The species has already been recorded from Egypt, Algeria, and Sudan, but also from Guinea, Senegal, Chad, the Central African Republic, Ivory Coast, Ghana, and Togo. The specimen illustrated here was caught west of Lomé in Togo.

Hemichromis letourneauxi

71

Hemichromis lifalili

LOISELLE, 1979

belongs to the most beautiful species of this genus. In addition to a warm brown colour on the head and the nape, the bodies of both sexes display ample reddish orange to fire-red shades. The gill-cover bears a prominent blotch partially bordered golden. The caudal peduncle does not have any markings. Only in the centre of the body there is a large dark blotch visible which is — as is the entire body, the head, and the fins — covered with numerous rows of small "iridescent" light blue speckles. This species attains approximately 10 cm in length with the females growing slightly smaller.

Distribution of *Hemichromis lifalili*

The

natural habitat

is the Congo River and its northern tributary, the Ubangi River, in Zaïre. *Hemichromis lifalili* could not yet be recorded from other regions.

Hemichromis lifalili — top: ♀, bottom: ♂

◆ *Hemichromis paynei*

LOISELLE, 1979

is also not a representative of the true "Red Jewel Fishes". Its body is almost exclusively coloured olive beige with the chin, throat, and parts of the lower belly being slightly lighter. Only during periods of courtship and spawning these parts become fire-red in the male. This colouration is then supplemented by small "iridescent" speckles on the head and sometimes also on the flanks. Depending on the disposition, both sexes may display a large dark blotch on the flanks and a narrow dark band may also become visible on the caudal peduncle. A dark gill-blotch bordered golden is always present. The small light blue speckles also appear in low and varying numbers in the

Distribution of *Hemichromis paynei*

Hemichromis paynei ♂

fins. The species attains lengths of approximately 10 cm. Distinguishing the sexes is extremely difficult and possible with certainty only when the courtship colouration is displayed.

The

natural habitat

of this species lies in the countries of Guinea, Sierra Leone, and Liberia. The fish inhabits flooding areas and rice-fields and, syntopically with *H. bimaculatus,* small current water-courses in savannas and forested areas. The specimen illustrated here was caught near Kambia in Sierra Leone where the species is widely distributed and occurs in all sorts of water-bodies. They are caught by the local people in weir-baskets already as juveniles of 5 to 6 cm in length to end their lives as tasty fish-soups. These fish are found especially frequently in richly vegetated water-courses. Since solid surfaces like stones are usually unavailable in these biotopes, the animals probably spawn on submerged branchwork or, more likely, on large leaves of aquatic plants.

Hemichromis paynei ♀

◆ *Hemichromis stellifer*

LOISELLE, 1979

has a beige-brown back and a rust-brown body. During courtship periods, this colouration changes into red in the male and orange-red in the female. Adult specimens often only show a dark gill-blotch whilst the central body-blotch is temporarily invisible. This species does not have a dark blotch on the caudal peduncle. The entire body including the fins is covered with various light blue speckles of up 2 mm in size. A total length of approximately 11 cm is reached.

The

natural habitat

extends from Gabon to Congo and Zaïre.

Care

These so-called "Jewel Fishes" permit a socialization with other Cichlids of approximately

Distribution of *Hemichromis stellifer*

their size as well as robust fishes of other genera. A spacious aquarium should be chosen with sizes of 130 cm in length, a depth of 50 cm, and a height of 40 cm being the mini-

Dense brush frequently shades the biotopes.

75

mum dimensions. Fine gravel with a grain-size of up 2 mm is recommendable as substrate. Rocks and roots should be placed directly onto the bottom plate and covered with gravel to form hiding-places and staying sites.

A richly planted environment is advisable for a stress-free husbandry. Since these fish do not damage plants, there are no restrictions regarding the decoration of the aquarium. One should however take their territorial behaviour into consideration and leave unplanted spaces in front of the hiding-places and caves which are the preferred sites to stay and to spawn. Otherwise the fish will arrange a space by themselves and contribute to their reputation to be plant-enemies. The richer the vegetation in the aquarium is, the more docile the fish become.

For most the species the water should be soft and acidic, but clean and rich in oxygen in any case. A slight movement of the water, initiated by a powerful pump, is advisable as well. The temperatures should be kept between 25 and 26 °C. A regular exchange of a quarter to a third of the tank-volume every week or fortnight should be obligatory.

Feeding "Jewel Fishes" is no matter of concern. Nutrious food like mosquito larvae, earthworms, enchytraeas, young fishes, and sometimes daphnia for a change, is required to obtain healthy strong specimens. Flake-food is also readily accepted.

The
breeding
is usually easy and very interesting to observe. Once a harmonic pair has bonded, you will soon be observing that a stone is being cleaned. "Jewel Fishes" belong to the substrate-breeders caring for their offspring in a parental family. The female usually lays her eggs on a stone where they are immediately fertilized by the male. 38 hours later, first signs of life become visible with embryonic fluids being pumped through the embryo becoming recognizable. At an incubation temperature of 25,5 °C the larvae hatch after 50 hours. Five and a half days later they swim free and are guided by both parents where the female however plays the major role. The male is mainly responsible for defending the territory which obviously increases in size with the juveniles growing and swarming.

During this time the aquarium has to be large enough and provide sufficient hiding facilities amongst plants into which other fishes may retreat. When it comes to safety for the offspring, the male may become highly aggressive.

Nauplii of the Brine Shrimp *Artemia salina* should be offered as first food. Powdered flaked-food is a valuable addition to the diet since the appetite of up to 300 young fish can hardly be satiated.

The swampy fields in Sierra Leone are the preferred habitats of *Hemichromis paynei*.

The

Genus Limbochromis

was suggested alongside with the genus *Parananochromis* by GREENWOOD in the 1987 revision "The genera of pelmatochromine fishes (Teleostei, Cichlidae): A phylogenetic review". It originally contained two species, whose systematic status was doubtful already at that time. The generic name *Limbochromis* also indicates that one was not too sure for the cladistic relationship with the other members of the *Pelmatochromis*-group of the past. Whilst the Latin word "limbo" in the generic name means that these fishes are "at the edge",

the English term "in limbo" rather indicates that both these species are "hovering" or "suspended". The genus *Limbochromis* differs from *Nanochromis* as well as from *Parananochromis* by a count of only 12 scales around the caudal peduncle instead of 13 to 14. In contrast to *Nanochromis* the upper lateral line does not run immediately below the base of the dorsal fin but is separated from it by at least one scale-row. Another important point of distinction supporting the special position of *Limbochromis* is found in the skeleton of the skull. Whilst the joint of the upper pharynx-bone in *Nanochromis* as well as *Parananochromis* is of the so-called *Haplochromis*-type, it is of *Tilapia*-type in *Limbochromis*.

This new species of *Hemichromis* was found in tributaries of the Kolente River at the border between Sierra Leone and Guinea.

The Birim River near Kibi in Ghana

Two interesting peculiarities of *Limbochromis* are found in the shape of the caudal fin, which obviously is lyreate in this genus, and in the geographical distribution. *Limbochromis robertsi* apparently lives outside the range of the members of the genus *Nanochromis*.

Since it appeared that, at the time of its description in 1987, insufficient preserved material was available, it was a challenge to import live specimens to obtain new information. In his paper titled "Vergleichende Unter-suchungen zur Systematik und Verwandt-schaftsproblematik ausgewählter Arten der Gattung *Limbochromis* und *Chromidotilapia* (Teleostei, Cichlidae)" which appeared in early summer of 1993, Lamboj transferred the species L. cavalliensis into the genus *Chromidotilapia* thus leaving *Limbochromis* monotypical.

◗ *Limbochromis robertsi*

(THYS & LOISELLE, 1971)

After the spectacular catch of a female in 1976 and the subsequent observations in captivity, it took almost another 15 years until ANTON LAMBOJ and his friends managed to re-record this fish again in its natural habitat. Besides the numerous females, now males were also caught and hence made available to aquarists. The reproduction was soon successful and the second generation already swims in the tanks today.

As far as the females are concerned, the species shows parallels with young *Chromidotilapia g. guntheri*. The males reach up to 11 cm in total length whilst the females are apparently already fully grown at 8 cm. Male specimens are very colourful with their heads being greyish brown in the upper parts. The gill-cover is of golden yellow colour and bears a very dark blotch which is bordered golden.

Distribution of *Limbochromis robertsi*

The iris of the eye is golden to beige. The back is coloured greyish brown and beige yellow right above the lateral line. Below the lateral

Limbochromis robertsi ♂

79

Limbochromis robertsi ♀

line, the beige turns into an iridescent yellow. The chest region is coloured yellow as well with the scales of the chest and belly having reddish violet edges. Whilst the pectoral fins are transparent and without colour, those of the ventral side have a faint orange colour and are bordered dark. The dorsal fin may be imperceptibly orange to sound yellow and is bordered dark red. Parallel to the edge there is a sound orange coloured stripe. Furthermore, the dorsal fin is covered all over with reddish violet spots. The anal fin is indistinctly orange with a dark bordering. A change to a faint violet with reddish violet spots is possible. In its upper part, the caudal fin is broadly bordered dark red with a parallel sound orange stripe. Two colour-stripes run — as if there were drawn with a ruler — as bordering lines of the dorsal fin up onto the caudal fin. The lower edge of the fin is black with a metallic white band. The central part of the fin is orange to sound yellow and has rows of reddish violet spots. In adult male specimens the shape of the caudal fin is lyreate. The ground colouration of females is generally darker with less yellow on the body and the fins. The dorsal fin has a bright yellow longitudinal band below a dark red bordering. The shape of the caudal fin is slightly subtruncate.

The

natural habitat

is probably restricted to the southwestern Kumasi area between the villages Kibi and Asiakwa in western central Ghana. The biotopes known from expeditions all lie approximately 5 km west of the small town of Kibi whose southern outskirts are crossed in eastern direction by the up to 4 metres broad and up to 2 metres deep Birim River. In its source area, the Birim runs through secondary forests and plantations on hills and fills a variety of small ponds in which no aquatic vegetation could be found. Emerse plants on the banks cover the water from sunlight. In August, the temperatures ranged around 22 °C whilst they measured some 25 °C in April. Adult males with territories of 3 m in length, 30 cm in breadth, and 10 cm in depth tolerated two or three large females as well as several fishes of moderate size in these areas. According to

THYS & LOISELLE, the females maintain several spawning sites and breeding caves in the large territory of a dominant male. Young males pretend to be females, participate in caring for the juveniles (?), and are therefore tolerated by the dominant male. The substrate consists of beige to brown gravel with a grain-size around 2 mm. Partially the ground is covered with dark grey lava-like stones of different size.

The area of my investigations was situated approximately 20 km north between Kibi and Asiakwa. It was a clear water-course of 1 to 2 metres in width and partially up to 40 cm in depth. The substrate was a light coloured sand with numerous rocks of up 30 cm in diameter scattered everywhere in the water. Also here, these were dark grey and lava-like. Aquatic plants were entirely absent. The water was very clean, clear, and had a slight current. The water-temperature measured 24 °C, the pH ranged around 6, and the conductivity was

established at 240 micro-Siemens at 24 °C. The stream was largely shaded by emerse plants of the banks. Fishes were numerous in

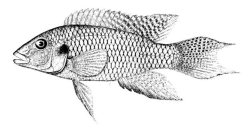

Limbochromis robertsi ♂

these water-courses, but the dense vegetation of the banks and the high number of rocks in the streams made successful catching almost impossible.

The rainforest is successively replaced by plantations.

Nanochromis sp. „Genema" ♀

The Genus Nanochromis

became quite voluminous by the tentative revision of the genus *Pelmatochromis* and its new arrangement in 1968. Whilst all its members reached a maximum of 9 cm in length and were considered small slender Cichlids until now, they were added also larger species. All representatives of this genus have 12 scales around the caudal peduncle, and both sexes, except *N. gabonicus,* have pointed ventral fins. In most of the species, the females have a metallic longitudinal band in the dorsal fin. The type species is *Nanochromis nudiceps.* An entirely new arrangement was suggested by the revision of the British ichthyologist P.H. GREENWOOD in 1987. According to his work, the larger growing species *N. cavalliensis* and *N. robertsi* from the western parts of Africa belonged to the new genus *Limbochromis* and were thus to be referred to as *L. cavalliensis* and *L. robertsi.* *Limbochromis cavalliensis* was however later transferred to the genus *Chromidotilapia* in

1993. Furthermore, the Cichlids of the central African area of Cameroon and Gabon were excluded from *Nanochromis* and placed in the new genus *Parananochromis.* According to GREENWOOD it contains *P. caudifasciatus, P. gabonicus,* and *P. longirostris.* Whilst *P. caudifasciatus* and *P. gabonicus* agree regarding habitus and reproductive biology, they both differ slightly from *P. longirostris.* The latter species rather resembles a mouthbreeder and carries the fry for a longer time in the mouth when it is to be transported to another site or in a case of danger. Similar observations were also made in another yet scientifically undescribed species of *Nanochromis.*

Hence, the original genus *Nanochromis* is divided into the new genera *Nanochromis, Parananochromis,* and *Limbochromis.* As to how long this status will be stable is to be watched. It is to be expected that further thorough studies may cause the transfer of other species into different genera and possibly even cause the necessity to further split the present genus *Nanochromis.*

▶ *Nanochromis consortus*

ROBERTS & STEWART, 1976

Distribution of *Nanochromis consortus*

 has not yet been imported alive. It is one of the species only described in 1976 in the paper "An Ecological and Systematic Survey of Fishes in the Rapids of the Lower Zaïre and Congo River". The reference specimens measured up to 52,2 mm in the males and 40,7 mm in the females. Unfortunately no description of the colour in life is available, but the preserved specimens already give a hint about the pattern and partially even about how these fish may look alive. A sufficient number of young and adult individuals were available for examination. The body appeared beige and lacks vertical bands. The individual scales are bordered with grey and the front part of the head as well as the nape are coloured rusty brown. Mature males and females differ in the colouration of the fins. In the areas of the soft rays of the dorsal fin and also in the two to three rays of the caudal fin the males show several vertically arranged rows of dark spots and sometimes also longitudinal stripes between the upper three rays. The lower rays are slightly darker. The ventral fins also tend to be somewhat more pointed in the males. In live specimens the upper and lower half of the caudal fin are coloured differently. The outer two to three rays of the anal fin are slightly darker, yellowish on the base, and have faint indications of vertical stripes. In the females, the dorsal, caudal, and hind part of the anal fins are uncoloured or partially slightly dark grey. Fully grown females in the collection have a black colouration of the anterior part of the anal fin which extends over a few rays. This is a colouration which is not found in males or females of any other known species of *Nanochromis* and thus represents a clear distinctive feature.

The

natural habitat

lies in the area of the lower Congo River near the village of Inga, approximately 60 km east of the town Matadi in Zaïre. The species is found in the southern tributaries and in calm zones of the main stream. The southern tributaries have water-temperatures between 21 and 24,8 °C, an oxygen content of 7,0 to 9,0 mg per litre, and a pH of between 6,5 and 9. At the time of examination in early August, the water-temperature measured 24,8 °C.

A rainforest water-course in Zaïre

One of the most beautiful small Cichlids is certainly

◆ *Nanochromis dimidiatus*
(PELLEGRIN, 1900)

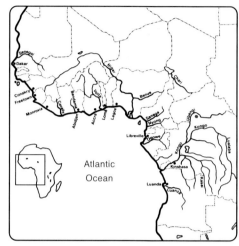

Distribution of *Nanochromis dimidiatus*

It is however presently still considered a rarity. Males reach approximately 8 cm and females about 6,5 cm in length. Clear points of distinction between the sexes are the large roundish blotch, sometimes added by several small spots, in the posterior spines or anterior soft rays of the dorsal fin in the female as well as a few bright silverish white scales forming a small blotch in the anal region. The female furthermore carries a metallic, bright, chromeyellow band in the spinous part of the dorsal fin. In order to draw a picture of this fish's colourfulness, I like to partially quote a description by DR. HERMANN MEINKEN: "This new species imposes in a nice shade of orange. In the slenderness of the body, *Nanochromis dimidiatus* largely resembles the well-known

Nanochromis dimidiatus ♂

N. nudiceps, but of course considerably differs by its entirely dissimilar colouration and by the snout being somewhat shorter. The colours change considerably. What was a specimen very dark in a state of excitement a few moments ago, it can be marbled all over the body or very light a few seconds later and may have a black lateral line, sometimes added by a row of black spots over the mid-body, or a short black band which extends just up to over the pectoral fins or no band at all. Generally, the ground-colour is an attractive greenish greyolive, the back is somewhat darker, the flanks are somewhat lighter, and the belly is whitish. The centres of the flanks are marked with a lateral band which also may be replaced by a black zig-zag band. The gill-covers bear beautiful golden or greenish golden spots. A great contrast is caused by the pretty orange red to tomato-red colouration of the lips, the lower part of the head, the throat, the front part of the belly, and sometimes even the lower part of the posterior caudal peduncle. The dorsal fin is greenish grey anteriorly with the rays being darker. Towards the edge, there is a bright lakegreen longitudinal band and the edge itself eventually has the colour of an orange. The soft-ray part of the dorsal fin is orange with dark brown spots arranged in transversal bands and a black rounded spot bordered light in the female. The pectoral fins are lake-green on the anterior rays and otherwise orange. The anal fin is greyish green anteriorly, becomes bright orange then, and has some spots arranged in transversal bands. The caudal fin is rounded, of orange colour, and there is a lake-green longitudinal line in the upper third of the fin in the male. Above the line, there are brown stripes or slightly irregular light and dark spots at the upper edge. The iris of the eye is blood-red. The beauty of females does not stand back for the males, but

Nanochromis dimidiatus ♀

it can be easily distinguished not only by the colour of the caudal fin, but also by the dorsal and anal fins being just pointed and not having long trails. Furthermore, they have an ocellate spot in the soft-ray area of the dorsal fin." The males may be quite aggressive if insufficient space is provided.

The

natural habitat

was indicated as Banghi, Ubanghi by G.A BOULENGER (1915). This is supposedly the frontier village of Banghi on the Ubangi River in the north of Zaïre on the border with the Central African Republic. Banghi is situated southeast of Kouango northeast of Bangui.

Care

Keeping this species is not without problems. Although the animals are quite small in size, they should not be kept in too small aquaria. The tank has to offer many hiding places, preferably heaps of rocks which provide small caves and have an obliquely arranged plate as a roof. The construction should be partly buried in the substrate for which fine gravel of up to 2 mm in grain-size is suitable. A very rich vegetation, turning the major part of the aquarium into an underwater jungle, is highly advisable.

Even if only a single pair is kept in a tank and there appears to be perfect harmony amongst the partners, one has to be prepared for bad surprises. Not only that the males fight each other viciously, they also may attack and seriously hurt their "beloved" females if these are not willing or able to spawn. Thus, hiding places very often make the difference between survival and death of a female.

Regarding the aforesaid, this species can be compared with *N. nudiceps* and *N. parilus*. It has repeatedly occurred that aquarists who thought they had a perfect pair and these precautions would be unnecessary, were taught a lesson when they spotted a dead female shortly after the pair had spawned.

The keeping of this species requires very clean water rich in oxygen. The water-tem-

peratures should range around 25 °C. A wide variety of live and flake-food should make up the diet.

If

breeding

is to be successful, soft acidic water is a precondition. The water in the breeding tank has to be rich in oxygen and poor in bacteria. A slight current is advisable. Spawning takes place in caves, usually on a roof-like ceiling of a cave. One may think of a mother-father family in the case of this species. The female cares for and maintains the eggs and larvae, and both parents together guide and guard the fry thereafter. The reproductive biology is comparable with that of the other *Nanochromis*-forms as well as with the species of *Pelvicachromis*.

Water-bodies are not always clear and clean. Cloudy waters muddy with clay are also biotopes for various species of Cichlids.

◗ *Nanochromis minor*
ROBERTS & STEWART, 1976

is yet to be imported alive. This species certainly belongs to the smallest representatives of
this genus. The reference specimens measured
only 23,8 mm in the male and up to 21 mm in
the female sex. In both sexes the tip of the
lower lip is coloured black. The chestnut-
brown colouration of the body intensifies from
bottom to top and becomes more spectacular.
On the upper edge of the eye, or above the eye,
there is a rust-red spot. The iris of the eye is
white. The folds of the gills and the chest are
coloured reddish in the male; the dorsal and
anal fins have dark brown spots. The upper
three rays of the caudal fin are coloured yellow
and have no spotted pattern. The subsequent
two to three rays below are marked with larger
brown blotches which transform into five to
six vertical bands downward. A faint lateral
band runs from the head over the flanks up to
the caudal peduncle. A violet blotch on the
body behind the pectoral fins which may
extend up onto the belly is characteristic for
the female. They too have a lateral band from
the head up to the caudal peduncle. Above and
below this band, a chestnut-brown band
extends up onto the central rays of the caudal
fin fading from the centre outwards. The rest
of the caudal fin is uncoloured. The dorsal and
anal fins are dark brown and bordered on the
edge.

The

natural habitat

ies in the area of the lower Congo River in
Zaïre. *Nanochromis minor* inhabits the rela-
ively calm ground levels between rocks in the
narrow side-streams of the huge Congo River.
At the time of examination the water was very
rich in oxygen. The substrate consisted of
muddy sand. Large boulders in the stream-bed
broke the torrent and their gaps and cavities
provided hiding-places and sites to stay. The
water-temperature on a site near the village
Tadi, 50 km down the river from Luozi, in July,
was established to be 24,9 °C, and 27,4 °C on a
site 5 kilometres east of Kinganga, below the

Distribution of *Nanochromis minor*

mouth of the river Grande-Pukusi, i.e. 25 km
farther east, at the same time of the year. The
oxygen-content was measured to be 8,0 mg
per litre whilst the pH ranged from 7 to 7,5.
The air-temperature varied between 23,3 and
30 °C. These figures are the average values
taken in the time from end of June to early
September 1973.

◆ *Nanochromis nudiceps*
(BOULENGER, 1899)

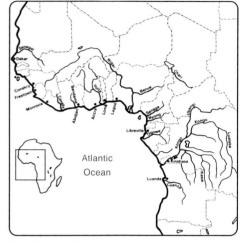

is a species long since known to aquarists, but a degree of uncertainty about its identity was observed in recent times. The fish illustrated as *N. nudiceps* in magazines and books during the past 30 years or so belongs, in the majority of cases, to a species described as *N. parilus* only in 1976.

"But how does *N. nudiceps* look then?" was the question which suddenly came up since only a few illustrations in fact showed the real *N. nudiceps*. The answer was quite simple since both species have almost the same appearance. Only very minor points of distinction exist. A slight difference is found in the colouration of the dorsal fin, but a major

Distribution of *Nanochromis nudiceps*

Nanochromis nudiceps ♂

feature is the additional pattern in the caudal fin. Reviewing all published photographs one finds that in fact both species are kept in captivity. The colour description published by the American ichthyologists ROBERTS and STEWART (1976) may help to determine *N. nudiceps*. Both sexes display a faint olive-brown on the nape and on the posterior part of the body. The cheeks are blue, the gill-folds reddish, a rusty brown spot lies above the upper edge of the eye, and the iris is dark. The dorsal fin and the upper part of the caudal fin are bordered dark.

The caudal as well as the anal fin are distinctly spotted. The ventral fins are whitish with the first spines being grey. Mature females display a violet zone in the belly area which may be more intense than in the males. The distant spines of the ventral fins are only feebly dark.

Further differences between the sexes are not conspicuous. Besides the morphological traits, *N. nudiceps* and *N. parilus* differ by the small spots which are arranged in vertical lines on the caudal fin and which are present in this form also in the species *N. splendens*. These remarks by ROBERTS & STEWART do thus not differ from the statements made by the English ichthyologist G.A. BOULENGER in his Catalogue of the Freshwater Fishes of Africa (1915). There the rows of spots are described to form vertical lines which are supported by a pattern.

The

natural habitat

was indicated to lie in upper Congo (?), Dolo (village or river?), and Stanley Pool (today referred to as Malebo Pool) by the English ichthyologist G.A. BOULENGER. This probably means the area of the lower eastern Congo River or its tributaries near the capital Kinshasa. Husbandry and reproduction do not differ from *N. parilus*.

After a successful catch the study of the habitats is extremely important. Extensive equipment is required.

◗ *Nanochromis parilus*
ROBERTS & STEWART, 1976

is the species aquarists have referred to as *Nanochromis nudiceps* for decades. Only once *N. parilus* was scientifically described in 1976 it was recognized how similar both these species are with regard to their colouration and pattern. Only a few features distinguish the species. *Nanochromis parilus* has a more contrasting pattern and a more conspicuous colouration in the upper bordering of the dorsal fin. An important point of distinction is the lower half of the caudal fin which usually lacks a pattern and is violet in colour. The male attains a length of up to 7 cm whilst the female is fully grown at 4,5 to 5 cm. An accurate determination of sex is only possible in adults. Whilst the males are longer and more slender, the females have a much more rounded belly and less produced posterior parts of the dorsal

Distribution of *Nanochromis parilus*

and anal fins. Females willing to spawn often evert their up to 2 mm long ovipositor already

Nanochromis parilus ♂

a couple of days before spawning actually begins.

The

natural habitat

of this species lies in the area of the lower Congo River near Wombe and the village of Inga in Zaïre. Here, the fish live in small side-streams of the main river, in the mouths of the tributaries or in calm zones of the River Zaïre as the Congo River is sometimes referred to as. These sites, which become small water holes in the dry season and are zones of calm water amongst the rocky banks of the main river, are the preferred environment of this species. In July, the water-temperature is approximately 28 °C, the oxygen-content of the water is 8,0 mg/l, and the pH ranges around 8. No data is available regarding total and carbonate hardness.

Care

Aquaria of at least 70 cm in length should be used for this species. Since these fish are often quite aggressive towards each other, a densely planted underwater-vegetation and well struc-tured rock arrangements are required to pro-vide the necessary hiding-places. In order to imitate the natural habitat, heaps of many moderately sized stones may be used whose gaps and crevices form ideal places for retreat.

The water should be soft and slightly alkaline. It has to be clean, clear, and rich in oxygen. A fairly strong current is recommend-able. The temperatures may vary between 25 and 26 °C. Fine gravel with a grain-size of up to 2 mm is appropriate. As live and flake-food are readily accepted, there are no problems with feeding although a variety of food is important.

Nanochromis parilus ♀

These recommendations also apply to

breeding.

Very few other fishes, no other Cichlids if possible, should be kept as company. Generally the "sires" of the species are lone-wolves and only a female ready and willing to spawn may change this attitude. After a very interesting courtship foreplay, the eggs are usually laid on the "roof" of a cave. Up to this point, a lot of patience is required from the keeper, and even if an apparently harmonizing pair has bonded, it is no guarantee for successful breeding. After spawning the female guards and cares for the eggs.

If disturbances are observed during the development of the eggs or if the water quality changes during this time, the female eats the eggs. Since the male lacks the understanding for such action, earnest quarrels may occur shortly thereafter. Insufficient hiding-places then may result in the death of the female. Even aquarists who thought they had a well

harmonizing pair and had successfully bred with these specimens in the past, were suddenly taught such a lesson.

If the development proceeds smoothly, the female leaves the cave with a school of small *N. parilus* after approximately eight days and both parents attentively guide their offspring through the aquarium and guard them. As is the case in all young fish, the fry should be fed the nauplii of the Brine Shrimp *Artemia salina* two to three times a day. Powdered flake-food completes the diet.

The water-temperature should be kept at approximately 28 °C during the development of the eggs and the first weeks thereafter. The water has to be very rich in oxygen and poor in bacteria. An exchange of the water is of extreme importance also during this time and contributes to healthy growth of the fry.

From the paper "An Ecological and Systematic Survey of Fishes in the Rapids of the Lower Zaïre or Congo River" by the American ichthyologists ROBERTS and STEWART in the

Nanochromis parilus ♂ (colour-variety)

year 1976, we know that we erred regarding the appearances of *N. nudiceps* and *N. parilus*. When the British ichthyologist G.A. BOULENGER described *N. nudiceps* in 1899, he referred to a specimen which neither had a black edged dorsal and caudal fin nor a light sub-edge. Besides the other minor traits of distinction the illustrations of BOULENGER introduced a fish which had only spotted markings in the caudal fin forming several vertical stripes. There was no black bordering or a light sub-edge. Only the form described as *N. parilus* by ROBERTS and STEWART in 1976 has this coloured bordering of the dorsal and the caudal fin. On the other hand, it lacks the spotted pattern in the caudal fin! The lower part of the caudal fin is transparent without any pattern.

However, our aquaria also housed fish where the males were of slightly different colouration having rows of spots in the lower part of the caudal fin. Aquarists thus thought this would now be the true *N. nudiceps*. The sup-

position had however to be abandoned again since offspring of *N. parilus* with the typical colouration included specimens with this colour abberation. This is however no appearance unique to captive keeping. According to ROBERTS & STEWART two different colour-morphs of *N. parilus* exist in the natural biotope as well. Are these possibly hybrids with *N. splendens,* or is this possibly another undescribed species? This presumption is opposed by the fact that the described features occur in the offspring of *N. parilus*. Even in pairs of perfectly typical *N. parilus,* male specimens are observed which have spots in the lower half of the caudal fin. It is however certain in any case that these two colourmorphs are definitely not *Nanochromis nudiceps* since this species lacks any banded or striped pattern in the dorsal and caudal fin.

The Ethiop River in Nigeria

◗ *Nanochromis* sp. "Bandundu"

Another spectacular Dwarf-cichlid which is yet to be scientifically described for the genus *Nanochromis* has been imported from Zaïre. The specimens were caught in the area of Bandundu. These fish resemble those which occasionally appeared in the pet-shops as *Nanochromis* sp. from Zaïre. The comparatively small growing animals — the females attain total lengths of approximately 4 cm, the males grow up to about 6,5 cm — are inhabitants of soft acidic waters. A feature of the mainly grey coloured species is the bright yellow colouration of the belly-region in the females. The males do not display any spectacular colour-features. It also appears that the latter lack the blue or red colour shades in the dorsal and anal fins.

Distribution of *Nanochromis* sp. "Bandundu"

Nanochromis sp. "Bandundu" ♀

◗ *Nanochromis* sp. "Genema"

A few specimens of these undoubtedly very pretty fish were imported in 1986 and originated from the area of Genema in the north of Zaïre. They became known as "Silver-blotched Nanochromis", but should rather be referred to as *Nanochromis* sp. "Genema". It is possible that this is the legendary *Nanochromis dimidiatus*.

Appearance and collecting locality make this presumption quite likely. Genema is situated only some 100 km south of Banghi, Ubanghi which is the type locality of *N. dimidiatus*. A final answer to this question will only be obtained by examination of preserved material in comparison with the type series which is kept in Paris. The 1986 imports were

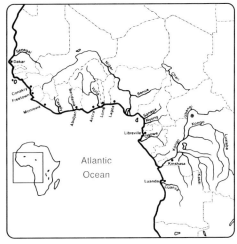

Distribution of *Nanochromis* sp. "Genema"

Nanochromis sp. "Genema" ♂

Nanochromis sp. "Genema" — front: ♀, rear: ♂

semi-adult specimens showing a slight deviation from the colouration as has been know to us for said species so far. For example, young females have a slightly greenish yellow colouration of the gill-cover. On reaching adulthood they however also showed the known reddish yellow colouration of the head. The green had vanished by then.

Unfortunately the husbandry and breeding of this form turned out to be problematic. Even when the water-quality was adjusted to their natural circumstances, i.e. very soft, very acidic water, their husbandry was still difficult. Already slight deviations of the pH toward neutral or alkaline levels caused severe health-problems and even became life-threatening after a very short time. Although the specimens spawned twice in my aquaria, no fry was obtained. The few other breedings by other keepers resulted in problems during the rearing period so that only a low number of specimens grew up and only very few pairs are available today.

It is to be hoped that these will form the basis of further successful breedings. The reproduction of this very pretty and small-growing — the males reach up to 8 cm in length, the females approximately 5 cm — Cichlid-species would be a real challenge for many Cichlid-fans.

◈ *Nanochromis* sp. "Kapou"

Another small and slender Cichlid was brought back from the area of Kapou by WOLFGANG HARZ in summer 1991. Kapou is situated 30 km southwest of Bangui in the Central African Republic. This small species of *Nanochromis* — the females grow up to approximately 4,5 cm, the males reach some 6,5 cm in length — has been also referred to as "Nanochromis sp. Zaïre-Red". These fish however lack the conspicuous red. The females of the "Kapou"-form resemble those of the forms "Zaïre", "Zaïre-red", and "Bandundu". They are hardly distinguished while the males are easily identified. Unfortunately the "Kapou"males lack any conspicuous colour-

Distribution of *Nanochromis* sp. "Kapou"

Nanochromis sp. "Kapou" — top: ♂, bottom: ♀

ation. They are rather drabs. According to a personal communication received from WOLF-GANG HARZ there are however specimens with red or blue coloured fins in the natural biotope. A specific trait appears to be the zig-zag band in both sexes. It clearly distinguishes this form from other but similar species in Zaïre and the Central African Republic. Studies, though incomplete, indicate that this might be the "true" *Nanochromis dimidiatus*. Should this presumption be confirmed, the discovery of their breeding behaviour would suggest that this species has never before been kept in captivity. The fish kept as supposed *N. dimidiatus* earlier thence would have belonged to other species.

I received some captive-bred specimens of this "Kapou"-form from PASCAL SEVERS in Zurich, Switzerland, in 1992. Approximately six months later the fish were mature. Unfortunately, many of the circulating Cichlids are contaminated with germs so that diagnoses and subsequent treatment is necessary to keep specimens healthy. At an age of approximately one year a pair was transferred into a richly vegetated breeding-tank. The animals are very attentive once they have spawned. Inspite of a neutral pH, spawn and larvae developed without major problems. In my case the water was poor in minerals and soft at a conductivity of 160 micro-Siemens. The water-temperature was only 24,5 °C. Despite the high pH and the relatively low temperatures unusual for *Nanochromis*-species the fry developed well. Whilst the female cared for the offspring in the cave, the male was permanently "on sentry" in front of it until the larvae swam freely. Thereafter the female leaves the fry. It does actually not appear that she is driven away by the juveniles. The male alone now continues guarding the about 80 juveniles. They stay closely together during the first days preferring places near the ground or amongst dense vegetation and are constantly guarded by the male. It was much a surprise to observe that the male almost permanently, hardly interrupted for short periods, carried a larger number of young fish in his mouth without that there was danger for the small ones. With the gill-covers spread wide and a greatly inflated gular sac it

carried up to 12 young fish for 10 minutes or longer eventually releasing them back into the school which meanwhile had stayed calmly. A short time later another batch was taken in the mouth and kept there for a while. This observation is extremely unusual in small West African Cichlids. A possible explanation could be that the juveniles are "cleaned", i.e. disinfected, by the oral phlegm. Other explanations have not yet been developed.

After approximately three weeks the juveniles become somewhat more independent and the school more loose. The male leaves the young *Nanochromis* more frequently and returns to the female. Provided favourable circumstances she will yet again have a clearly recognizable swollen belly and set up new spawn. By frequently proceeding into caves they both show their willingness to spawn again. In one case this even happened right after an extensive exchange of water at a temperature of only 22 °C. Spawning took almost two hours. Nothing similar to the female "courtship-dance" of e.g. species of *Pelvicachromis* could be observed. The eggs are laid on the roof of the cave and partially hang down from the ceiling of the hiding-place on an "egg-thread". Being a cave-spawning species and forming a mother-father-family, the female takes care of the eggs and larvae. Once these swim free, the male takes over to guide and protect the small *Nanochromis*. These observations are entirely in contrast to the reproductive behaviour recorded for the forms *Nanochromis* sp. "Zaïre", "Zaïre-Red", and "Bandundu". It is therefore likely that they are different species. It cannot yet be excluded that this form is identical with *Nanochromis* sp. "Bandundu". On the other hand Bandundu is a few hundred kilometres more south — a big distance for a small fish.

Another species was imported already in 1985 and is referred to as

◆ *Nanochromis* sp. "Kindu"

and was portrayed elsewhere as *Nanochromis* sp. aff. *minor*. According to my investigations only females have been imported so far, thus making it impossible to breed with this species. *Nanochromis* sp. "Kisangani" may, in my opinion, be the same form differing only in some very minor points in colouration. Notwithstanding this, it would be sensible to refer to both varieties separately by indicating the origin, i.e. *Nanochromis* sp. "Kisangani" and *Nanochromis* sp. "Kindu". Any other marketing name, rather meant to increase sales, should not be accepted.

Distribution of *Nanochromis* sp. "Kindu"

Nanochromis sp. "Kindu" ♀

99

One of the most spectacular recent imports is a fish sometimes referred to as *Nanochromis squamiceps,* a fish which should rather be named

▶ *Nanochromis* sp. "Kisangani"

until its systematic position is established.

These fish belong to the small Cichlids which are conspicuous for their attractive colouration. The females show a bright silverish area of scales on the flanks as their typical sexual feature. It extends from the posterior back over the upper belly up to the anterior parts of the lower tail and ends in a iridescent silverish anal-blotch. The ground-colour of the body is generally some sort of grey to reddish brown. Depending on the disposition of a specimen this colouration may partly change into a sound red with the belly usually remaining soft rosy. The adult size of a female is approximately 5,5 cm in total length whereas males may grow up to 8 cm. The latter are lacking the

Distribution of *Nanochromis* sp. "Kisangani"

prominent silverish scales. Whilst females have an unpatterned caudal fin, several distinct horizontal rows of spots are recognizable in the males forming 7 to 8 vertical stripes.

Nanochromis sp. "Kisangani" — top: ♂, bottom: ♀

The
natural habitat

of these "Silver-bellied *Nanochromis*", as they are sometimes referred to as, lies in the area of Kisangani in Zaïre. The specimens presently available were probably caught by HEIKO BLEHER and imported alive in low numbers.

In the wild, the fish inhabits very soft acidic waters, but husbandry experiments show that healthy keeping is also possible in slightly alkaline water of moderate hardness. When it however comes to breeding the natural water-values are of extreme importance for the development of the eggs and larvae.

Care

This species represents a real enhancement of the variety of fishes for the aquarium. It is a timid bottom-dweller which requires a not too small, partly richly vegetated aquarium with many small caves if its keeping is to be optimal. A good water-quality is of outstanding importance which means that a third to the half of the tank-volume is to be exchanged regularly, i.e. weekly if possible.

For the
breeding

in caves, old, well watered tubes of bamboo with one end open have proven very successful. They are usually given preference by the fish before coconut-shells or the not so fancy flower-pots. *Nanochromis* sp. "Kisangani" is fairly productive and broods of 100 young fish are to be considered the average. This species belongs, as all presently known species of *Nanochromis* and *Pelvicachromis* from "West Africa" do, to the cave-spawning Cichlids which care for their offspring in a mother-father-family. This means the female almost exclusively takes care of the eggs and the larvae whilst the male usually defends the breeding territory, sometimes with the assistance from the female. The fry swims free after approximately eight days, provided appropriate values and a good quality of the water and temperatures around 27 °C. Thereafter both parents guide and guard the offspring.

Nanochromis sp. "Kisangani" ♂

◆ *Nanochromis* sp. "Zaïre"

is yet another small-growing undescribed species awaiting a scientific name. It has been imported to Europe for the first time in 1978/79 and was mistaken by the aquarists for *N. dimidiatus* for quite some time. Its appearance, size, colouration, and pattern indicate a close relationship to the species *N. consortus, N. minor,* and *N. splendens* described by the American ichthyologists ROBERTS and STEWART. It is however different in colour and pattern and in some morphological details. The 1976 paper of the aforementioned authors contains a remark which might refer to this undescribed species. Two specimens in the British Museum are mentioned which very much agree with the description of *N. dimidiatus* but differ from the species described as new. They were however not examined in detail since they are in a bad state of preservation. It was how-

Distribution of *Nanochromis* sp. "Zaïre"

ever clear that these "*Nanochromis* sp. undet.", as they were referred to as, which were caught

Nanochromis sp. "Zaïre" ♂

together with the new species in Zaïre, belonged to a new species closely allied to *N. dimidiatus*.

Nanochromis sp. "Zaïre" attains a length of approximately 6,5 cm with the females being fully grown at 3,5 to 4 cm. They are very timid fish with very little intraspecific aggression. The sexes are distinguished by the different adult sizes, the females have a metallic longitudinal band in the dorsal fin, without or with only a very faint spotted pattern in the caudal fin, slightly shorter ventral fins, and usually a rounded belly.

The

natural habitat

of this species lies — according to the present state of knowledge in the lower Congo River and its tributaries in Zaïre.

Care

Keeping these fish is relatively easy. Despite the relatively small size, aquaria with a minimum length of 70 cm should be made available with even larger tanks being even more suitable. This small species is conditionally adequate for being kept together with larger Cichlids.

If they are kept in a tank of 130 cm, or even better 160 cm in length, a pair of a species of *Chromidotilapia* and a *Pelvicachromis*-form can be kept with them without problems. The available ground space is always the crucial factor. The decoration, i.e. the number of available caves amongst rock constructions and underneath coconut-shells, the number of plants, which turn the tank into a densely vegetated jungle with many places to hide and stay, usually allows the keeping of several bottom-dwelling species of Cichlids at the same

Nanochromis sp. "Zaïre" ♀

time. This species also definitely requires clean water rich in oxygen. It may be of a soft to moderately hard, slightly alkaline quality.

The

breeding

was successful in moderately hard, slightly alkaline water. For this specific purpose the animals may be housed in a smaller tank which is however to be decorated the same as the main aquarium.

This species spawns — as all representatives of the genus do — in a cave, mostly on the ceiling of it. The courtship is quite vigorous. Thereafter, the female cares for the fry whilst the males defends the breeding territory. After approximately eight days the females leaves the cave guiding a small number of young fish together with the male through the vicinity of the breeding cave. The number of juveniles is small with 30 to 40 being the average. In this case, the first food should also consist of the nauplii of the Brine Shrimp alternating with powdered flake-food and should be given three to four times a day. To hide the young fish from nocturnal predators they are brought back into the cave by the parents every evening where they spend the night together with the mother. The father stays outside the cave at a secluded spot.

This procedure can however only be observed when "day" and "night" change at fixed times. It is therefore recommendable to employ an automatic timer set up for a regular 12 hour-rhythm.

Examination of the water sometimes has surprising results.

The name *Nanochromis* sp. "Zaïre" was applied to another small-growing Cichlid-species which shall be referred to here as

♦ *Nanochromis* sp. "Zaïre-Red"

because of its colouration and to allow distinction. Both species reach lengths of approximately 6,5 cm in the male and 4,5 cm in the female. The females of both resemble each other very much and are not easily distinguished. The males however differ considerably in their colourations, especially with regard to the dorsal fin and the upper part of the caudal fin and are thus easily identified. Whilst the specimens of the species portrayed in the previous chapter have a grey to light blue colouration, the respective zones in this new species are sound red. It is fairly unlikely that this is just a local variety with a different colouration of the same species. As to how far

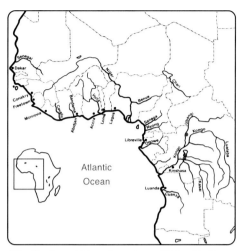

Distribution of *Nanochromis* sp. "Zaïre-Red"

both these species may be allied to *Nanochromis* sp. "Bandundu" is yet to be investigated.

Nanochromis sp. "Zaïre-Red" ♂

Nanochromis sp. "Zaïre-Red" has been propagated to a still limited extent by successful captive breeding. It is to be presumed that this beautiful timid fish will soon be a fixed asset of the aquarists. Since no data on the natural habitat was available for this *Nanochromis*-species, its keeping and especially its breeding was an enormous problem. It was just known that the animals originated from Zaïre. Observations however soon revealed parallels to other species of the genus *Nanochromis*. Aquaria decorated partly with densely growing plants and rock constructions providing many caves and hiding-places have also in this case proven optimal. A high quality of the water turned out to be very important which meant frequent partial exchanges of the water. Kept together with other small unaggressive species in moderately hard, slightly alkaline water its husbandry is relatively easy.

The

breeding

of these small West Africans is unfortunately not hassle-free. The quality of the water is of outstanding importance. Very soft, acidic conditions with a pH of 5,0 to 5,5 are advantageous for a successful development of the fry. *Nanochromis* sp. "Zaïre-Red" is a cave-spawning Cichlid which prefers small cavities for spawning. Halved coconut-shells with a small entrance or old bamboo-tubes with an opening on one side provide ideal places.

The species reproduces in the form of a mother-father-family with the female exclusively caring for the eggs and the larvae. During this time, the male guards and defends the breeding-territory. The eggs are laid on the ceiling of the chosen cave. At a water-temperature of 26 °C on average the embryos emerge

Nanochromis sp. "Zaïre-Red" — top: ♂, bottom: ♀

from the eggs after approximately 72 hours and subsequently lie on the ground of the cave. Whilst the development proceeds, the larvae are "hung" on walls or the ceiling again by the caring female. Another eight days later they can swim and are usually guided out of the breeding-cave for the first time in the early morning hours. It is commonly observed that the female collects the young ones in the mouth and brings them back into the cave after a short excursion. Now the male also joins in with the care of the offspring. Since the small *Nanochromis* do not stick to a school during the first days, both parental specimens catch individual runaways with the mouth and bring them back to the school which almost exclusively stays near the ground. The young fish are still relatively small and thus require appropriately sized food. Even newly hatched *Artemia salina* are usually too big as first prey

and offering these would just result in the juveniles dying of starvation. However, very small Brine Shrimp are available. If the first critical days are survived, the small *Nanochromis* grow quite rapidly if provided a variety of food in good quantities and a healthy water-quality. They may reach maturity after approximately six months. Some 40 to 50 descendants can be obtained from one breeding cycle.

Despite "antagonistic fishes" in the breeding tank, the harmony between the pair is not necessarily always optimal and slight disagreements may soon lead to excessive bitings. Since the devotion for the fry decreases already after a few days, this may result in losses of young fish on a large scale. The alternative of rearing the larvae "artificially" is however also not without its problems and may frequently result in a larger loss.

A biotope for species of *Chromidotilapia* and *Pelvicachromis* in southern Cameroon

◆ *Nanochromis splendens*
ROBERTS & STEWART, 1976

is a species which has not yet been imported alive. It belongs to the small species of the lower Congo River. The type specimens have lengths of 4,6 cm. The maximum total length is not yet known, but should be around 7 cm.

The colouration of these fish is interesting. In both sexes the gill-areas are canary-yellow anteriorly and orange-red in the posterior part. The eyes have a red spot on the upper edge and the iris is white. This colouration is very conspicuous and lets the eyes appear larger than they actually are. The scales are bordered grey. Five or six indistinct vertical bands cross the body. A shapeless blotch, sometimes becoming dark blueish violet, ornaments the gill-covers at the height of the pectoral fins. The cheeks of the male are partly blue and the gill-folds red, the spinous part of the dorsal fin is brown with the tips of the spines being yellow and the soft-ray part being bordered dark grey. The upper half of the caudal fin is coloured yellowish orange and contains several straight lines. The lower half of the caudal fin has seven to eight vertical lines of carmine-red colour and is otherwise transparent. A part of the anal fin is coloured similarly. Occasionally, the uncoloured parts of the fin may display shades of grey. The long spines of the ventral fins are black whereas the soft rays are white.

In females, the cheeks are canary-yellow, the gill-folds are dark yellowish, and the belly is violet. The spinal part of the dorsal fin is reddish orange up to the base and bordered yellow. The caudal fin is yellowish, reddish on the caudal peduncle, and bordered. The anal fin is uniformly blueish. The spines of the ventral fins are bright white whereas the soft rays are red.

The

natural habitat

was indicated as the lower Congo River between the villages Tadi, Kinganga, and Inga in Zaire.

Distribution of *Nanochromis splendens*

◆ *Nanochromis squamiceps*
(BOULENGER, 1902)

has also not yet been imported alive. A single specimen served for its description in 1902 and its validity has been questioned ever since. A confusion with *Nanochromis* sp. aff. *caudifasciatus* by the English ichthyologist G. A. BOULENGER also haunts through various publications. The American scientists ROBERTS and STEWART again reported about *N. squamiceps* in 1976.

Unfortunately to date only very little is known about the colouration of these fish. Their body-shape resembles that of *N. dimidiatus*. The upper part of the body is light brown whereas the lower part is white. A dark horizontal stripe runs on either side from the eye to the edge of the gill-cover. The soft sections of the dorsal and anal fin and the caudal fin have numerous transversal rows of dark spots. The species has large scales in the nape and on the cheeks. Scales of this size are very seldom in species of *Nanochromis* and have led to its scientific name.

The

natural habitat

was indicated as Lindi River, upper Congo.

▶ *Nanochromis transvestitus*
ROBERTS & STEWART, 1984

Typical for most aquarium fishes is the fact that the male specimens have a more beautiful colouration and often a more brilliant pattern. This is however entirely different in a West African Cichlid which has been imported for the first time in the mid-eighties. Here, the roles are swapped with the female having the prettier colouration and the more attractive pattern. This was the reason for the American ichthyologists ROBERTS and STEWART to name this species *"transvestitus"* in their paper published in 1984. The husbandry of this form is unfortunately problematic.

The

natural habitat

of *Nanochromis transvestitus* was discovered in the Lake Mai-ndombe during an ichthyological expedition through the area of the

Distribution of *Nanochromis transvestitus*

Zaïre Basin in 1973. The Lake Mai-ndombe is also known as Inongor or Lake Leopold II. In the south, it drains into the River Fimi, also referred to as Lukeni, which is connected to

Nanochromis transvestitus ♂

the Kasai, one of the major tributaries to the mighty Zaïre River.

The ecological details of the lake are of interest. The Mai-ndombe has a sandy ground. Where it is not covered with rocks, there is a tight-baked sediment which is usually overlaid with remains of plants. The shore areas usually consist of permeable rock covered with the branchwork of partly dead trees. The water is almost "black", i.e. has the colour of very strong tea, and is distinctly acidic; a pH of 4,0 was established. This extreme and unusually low value is obviously the most important factor for a successful keeping in captivity.

According to the sparse experiences made so far,

Care

keeping *Nanochromis transvestitus* and breeding it is very problematic if the natural water-quality is not provided. This means that soft water of good quality is a precondition for a successful keeping. As is the case in all other species of this genus, *N. transvestitus* is mainly a bottom-dweller. Fine gravel or sand should therefore used as substrate. Small caves and rock constructions as well as partially dense zones of aquatic plants should be provided.

Nanochromis transvestitus is a concealed breeder often inhabiting caves.

Nanochromis transvestitus is a cave-breeder forming a mother-father-relationship in which the female cares for the eggs and larvae and the male guards and protects the spawning territory. The juveniles are alternatively guided and defended by either parent. The sharing of the duties is clearly defined and does not differ from other related species. It is however noteworthy that the male is more active in comparison and that the female often takes over to protect the territory.

At the time of "sunset", the school of young fish — approximately 60 on average — is brought back into a protective cave, and it can be observed that both parents alternatingly are "on sentry" inside. The young fish readily consume newly hatched nauplii of *Artemia salina* as first food.

It is obviously of importance for successful breeding and the development of the fry to provide very soft water with a pH of between 4 and 5. It should however be remarked that my specimens reproduced as well in water with a pH of 5,8 and the fry developed without complications. Changes of the pH are, on the other hand, very dangerous for the offspring during the first days of life. To fortify the water with the necessary acid, filtering over peat is advisable.

Nanochromis transvestitus certainly belongs to the most spectacular discoveries in the past few years. Because of their requirements regarding a very special water-quality, these West Africans unfortunately appear however to be fish for the "specialists". It is to be hoped that this may change over the generations bred in captivity.

Nanochromis transvestitus ♀

The

Genus Parananochromis

was one of the new genera described by GREENWOOD in 1987. It contains the somewhat larger species which all share the feature of a high body. Notwithstanding this, the species of *Parananochromis* and *Nanochromis* share many traits and may be regarded as sister-genera. An important point of distinction is the course of the upper lateral line which lies immediately below the dorsal fin in *Nano-chromis*, but is separated from it by one or one and a half scales in *Parananochromis*. Both *Parananochromis* and *Nanochromis* have 12 scales around the caudal peduncle and thus differ in this trait from the genera *Limbochromis*, *Chromidotilapia*, and *Pelvicachromis*.

Just five species presently form the genus *Parananochromis*:
Parananochromis longirostris
Parananochromis caudifasciatus
Parananochromis gabonicus
Parananochromis sp. "Rio Muni"
Parananochromis sp. "Belinga"

Parananochromis caudifasciatus — top: ♂, bottom: ♀

112

A very rare species of *Parananochromis* was made available by "travelling aquarists". It is

▶ *Parananochromis caudifasciatus*
(BOULENGER, 1913)

which was caught and imported alive to Europe by the group of enthusiasts around the aquarist OTTO GARTNER of Vienna, Austria, in 1975. This made observations in captivity possible for the first time.

The males of this species may reach 11 cm in length whilst the females are fully grown at approximately 8 cm already. The sexes can be easily distinguished. Female specimens have a bright metallic band in the spinal part of the

Distribution of *Parananochromis caudifasciatus*

Parananochromis caudifasciatus ♂

dorsal fin and sometimes display a somewhat darkened area in the anal region behind the belly.

The

natural habitat

of this species lies in the tributaries of the Nyong River in southern Cameroon. In 1975, the fish was caught near Pouma, a small village on the road from Edea to Jaounde. However, four years later OTTO GARTNER and myself visited this site again, but could not find any specimens. Approximately one hundred kilometres south, in a tributary to the Soo River, a branch of the River Nyong, a few kilometres outside of Sangmelima towards Ebolowa, we caught five large specimens in

the thicket of aquatic plants in a pond-like water-body. The colouration right after the catch was very pretty, but was unfortunately never again displayed in captivity. A sound golden yellow with a contrasting violet-blue was prominent.

In the aquarium, the wild-caught specimens remain shy and faded. At this site, the water was clear and of a brown colour. The ground consisted of fine gravel containing loam. The water was partially quite deep and exposed to direct sunlight for a longer time during the day. At noon, an intensity of 50 000 Lux was measured, but nonetheless was the temperature of the water established to be only 22 °C. Further localities are in the southern Dja River-system in southern Cameroon and in the neighbouring Rio Muni.

Parananochromis caudifasciatus ♀

◗ *Parananochromis gabonicus*

(TREWAVAS, 1975)

Distribution of *Parananochromis gabonicus*

The only known specimen which served for the scientific description of the species was caught during an expedition in 1957. Nothing was understood about the colouration in life, the behaviour, and the reproductive biology for many years. Apparently for the first time, specimens were brought back alive from an expedition to Ghana by Messrs. LANDSBERG, NUMRICH, and WUNDERLICH in 1986. It is a highbacked, somewhat larger growing cave-spawning Cichlid with a father-mother-family. Whilst adult males may measure around 12 cm in total length, female specimens are fully grown at approximately 10 cm. A typical feature of the female is a deep orange to yellow

Parananochromis gabonicus ♀

115

Parananochromis gabonicus ♂

iridescent band in the spinous section of the dorsal fin. It is characteristic for the female in all species of this genus except for *P. longiros- tris*. *Parananochromis gabonicus* have a high degree of intraspecific aggression, and the husbandry thus requires appropriately large, well structured aquaria. For breeding, the natural water-values are to be taken into consideration, i. e. very soft water poor in minerals is to be provided. Since these fish are not especially colourful or attractively patterned, the demand is not specifically high and although they have been bred several times, this very interesting species may have meanwhile become extinct in the aquaria again.

The

natural habitat

lies in Gabon. The type specimen was caught in a water-hole next to the road from Mitzic to Medounen in the area of the Okano River. In a wider sense this is a tributary to the Ogowe running closely to the frontier with Rio Muni. According to LANDSBERG, NUMRICH, and WUNDERLICH, further localities exist along the roads connecting Oyem, Momo, and Minvoul on the upper courses of the rivers Abanga and Okano, along the frontier of Rio Muni, and in the buttresses of the Crystal Mountains in the west of the country near Edoum.

◗ *Parananochromis longirostris*
(BOULENGER, 1903)

has been imported alive for the first time only a few years ago and studies are still far from being complete. It is a comparatively large species with the males reaching approximately 13 cm in length and the females 11 cm. The sex of adult specimens is easily determined by the males having a lobed extension of the caudal fin between the sixth and eighth upper ray. The appendage is only indistinctly recognizable in the female. The posterior zones of the dorsal and anal fins of the males are more produced longer and show, as the caudal fin does, a more conspicuous pattern of light speckles.

On occasion of collecting trips by NUM-RICH, LANDSBERG, and WUNDERLICH in 1986 and 1987, the fish was circumstantially observed in its natural habitat and subsequently imported alive for the first time. *Parananochromis longirostris* inhabits small

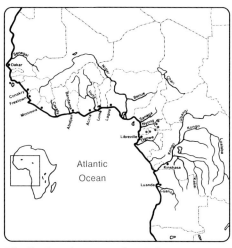

Distribution of *Parananochromis longirostris*

water-courses as well as large rivers. The animals were repeatedly observed to form small groups of 10 to 12 specimens above extensive

Parananochromis longirostris ♀

117

areas of bare sand ground without any cover. The presumption that this species thus might be a mouthbrooder and not, as typical for the genus, a substratum spawner, was however incorrect. Although the animals take the fry into the mouth for longer periods in cases of danger or for transport, this is definitely no

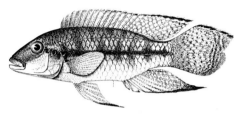

Parananochromis longirostris ♀

mouthbrooding behaviour in the true sense. The breeding behaviour may possibly depend furthermore on the inhabited biotope or the presence of predators and the decoration of the aquarium.

The strategy of mouthbrooding certainly has a protective function, although it is yet to be investigated whether a disinfectant effect is obtained by the oral phlegm since a similar behaviour is also observed in the species *Nanochromis* sp. "Kapou".

Parananochromis longirostris is the type species of the genus *Parananochromis* established by GREENWOOD in 1987. As it differs considerably from all other species assigned to this genus, another revision would be urgently required.

The
natural habitat

is indicated in the description as Kribi and Ja Rivers in southern Cameroon. Despite various attempts I did not manage to record this species again from the vast area of southwestern Cameroon up to the southern border with Rio Muni. It possibly no longer exist in the Kienke River, formerly named Kribi. These fish can only be found farther in the east, in the Dja River-system, formerly referred to as Ja. Their biotopes extend to the south up to central Gabon and to the southwest up to Rio Muni. In this region they inhabit watercourses with very clean, soft, slightly to strongly acidic water with a slow current. NUMRICH furthermore recorded some localities in Gabon, i.e. Ntem (Campo), Benito (Woleu), Invindo and the tributaries Djadi and Djuja, Liboumba, M'Passa, and Sébé. It is interesting that no records have been made from the Ogoowe so far.

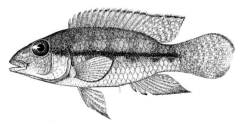

Parananochromis longirostris ♂

Washing site and biotope

Parananochromis sp. "Rio Muni"

has apparently not yet been imported alive. This fish shares many features with the species *P. caudifasciatus*. The males probably reach total lengths of 8 cm with the females growing only slightly smaller. Both sexes display a dark blotch on the gill-cover from where a faint lateral band extends up to the root of the tail in the female. The dorsal, caudal, and anal fins lack all pattern in the female, but not in the male. There, the soft-ray area of the dorsal and the anal fin and the entire caudal fin are patterned with small, elongate, dark speckles which form several vertical stripes. Both sexes have produced ventral fins.

The

natural habitat

lies in Rio Muni, the former Equatorial Guinea or Spanish Guinea between Cameroon and Gabon on the Atlantic Ocean. The species inhabits small water-bodies and ponds as well

Distribution of *Parananochromis* sp. "Rio Muni"

as slowly current water-courses of the southern tributaries to the Ntem, the frontier river between Rio Muni and Cameroon, respectively between Gabon and Cameroon.

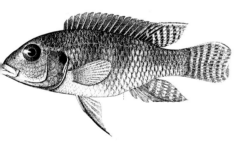

Parananochromis sp. "Rio Muni" ♂

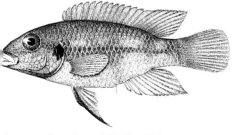

Parananochromis sp. "Rio Muni" ♀

Small flowing water-courses with a vegetation of species of *Anubias* are the preferred localities for many Cichlids.

119

The slender, scientifically undescribed

▶ *Parananochromis* sp. "Belinga"

was discovered and caught during an excursion by KLAUS LANDSBERG, JÖRG WUNDERLICH, and ROLAND NUMRICH in early 1987. It occurs in the region of the Belinga settlement, northeast of Makokou on the Ivindo River in the West African country of Gabon. The site is a small fast flowing stream in the vicinity of the village of Mayebout "which originates from the mountains and mouthes into the Ivindo already after a few kilometres".

According to data compiled by the participants of this trip, the water is relatively cool at 22,9 °C. By having only 22 micro-Siemens it

Distribution of *Parananochromis* sp. "Belinga"

Parananochromis sp. "Belinga" ♂

Parananochromis sp. "Belinga" — top: ♀, bottom: ♂

turned out to be very poor in minerals, and with a pH of 4,6 very acidic. This fish, presently dealt with as *Parananochromis* sp. "Belinga", was imported in a few specimens in 1987 and bred in small numbers since then. Adult males attain total lengths of approximately 9,5 cm, whilst females are fully grown at some 8 cm. Provided soft and slightly acidic water in appropriately sized, well filtered tanks, the husbandry is easy. The breeding is

however much more of a problem and probably depends on the natural reproduction seasons. *Parananochromis* sp. "Belinga" is a cave-spawner with a father-mother-family structure. Husbandry and breeding conditions are comparable with species of *Pelvicachromis* and *Nanochromis*. The so far quite few successful breedings are still a challenge for the enthusiasts.

121

The

Genus Pelvicachromis

is composed of small-growing, usually slender Cichlids which inhabit clear, slightly current waters rich in oxygen. The majority of species are colourful. Compared to the males, the females grow smaller and are recognized by a bright metallic band in the dorsal fin. During periods of courtship, when the colours are displayed to the full extent, they can be easily distinguished by their sound red, in some forms brilliant blue bellies highly contrasting with the yellowish golden ground-colour of the bodies.

A typical feature of this genus is the rounded or flattened tip of the ventral fins in the females distinguishing them clearly from other small West African Cichlids like those of the genus *Nanochromis*. All species of this genus have, as most Cichlids do, a count of 16 scales around the caudal peduncle. The shape of the head also differs from other Cichlids by the front portion of the head and the nasal area being unmistakably curved whilst the lower line of the head, i.e. the chin area, is almost straight.

All members of this genus belong to the cave- or substratum-spawning Cichlids. The breeding behaviour follows a mother-father-structure where the female cares for the fry and the male protects the breeding territory during the development of the offspring. Both parents guide and guard the fry with the female usually being the more active partner.

Presently eight species form the genus *Pelvicachromis* which furthermore includes a high number of colour varieties:

1. *Pelvicachromis* sp. "Guinea"
2. *Pelvicachromis humilis*
3. *Pelvicachromis pulcher*
4. *Pelvicachromis* sp. aff. *pulcher*
   from Nigeria
5. *Pelvicachromis roloffi*
6. *Pelvicachromis subocellatus*
7. *Pelvicachromis* sp. aff. *subocellatus*
   from Nigeria

8. *Pelvicachromis taeniatus*

Fortunately, all species, including the undescribed one from the Bandi River in Guinea, are known as live animals and have been studied in captivity.

The type species of the genus is *Pelvicachromis pulcher*. Based on the habitus, the genus can be arranged in four groups:
1. the *"pulcher"*-group
   consisting of elongate slender species
2. the *"humilis"*-group
   consisting of elongate species with a higher body
3. the *"taeniatus"*-group
   containing small-growing slender species and
4. the *"roloffi"*-group
   which is assembled of small-growing species with a higher body.

The rare Water-lily *Nymphaea daubenyana* from the Ethiop River.

◗ *Pelvicachromis humilis*

(BOULENGER 1916)

Colour-morph "Kenema"

has rarely been imported and studied only in a very few instances. Mr. ROLOFF of Cologne, Germany, caught these fish on numerous occasions in Sierra Leone and managed to bring them back alive. The males reach up to 12,5 cm in total length whilst the females are fully grown at approximately 10 cm. Both sexes have no or only very indistinct transversal lines over the body and a grey to yellow coloured gill-region. The gill-covers are marked with a dark blotch. The dorsal and anal fins of the male are pointed and slightly rounded in the female. Female specimens display a bright light zone in the dorsal fin; in contrast to the male their ventral fins are rounded.

Distribution of *Pelvicachromis humilis* Colour-morph "Kenema"

Pelvicachromis humilis Colour-morph "Kenema" ♂

123

The
natural habitat

lies in Liberia, the southeast of Guinea, but mainly in Sierra Leone, i.e. in the vicinity of the town Kluema and the areas around Koribundu and Pujehum in the south of the latter mentioned country. The fish inhabits gently flowing water-courses with very soft, acidic water of a pH under 6 and a conductivity of 30 micro-Siemens at 26°C water-temperature. These are not always forested areas, but also plantations and bushlands where a partial lack of shade may result in an increase of the water-temperatures by 2 or 3°C.

The
husbandry

of this species requires tanks with a base-size of 130 by 50 cm or larger. The height is of secondary importance, but should not be less than 40 cm although this species is a bottom-dweller occupying mainly the lower levels. If the tank is too shallow and too few floating plants cover the surface, the fish become very shy and nervous. The substrate should consist of dark gravel with a grain-size of up to 3 mm in diameter. Caves should be created by piles of rocks in which only rounded and calcium-free stones should be chosen. This should be tested with a few drops of acid before they are used for decoration. Rock constructions are placed directly onto the bottomplate and then partially covered with gravel. Bog-oak provides further items for an attractively decorated aquarium. Small-growing robust plants such as Anubias nana and the Congo-water-fern *Bolbitis heudelotii* may be clamped in between rocks or attached to the bog-oak. Two or three large *Nymphaea lotus,* in green and red, should complete the decoration except for some other plants. A powerful filter-system resulting in a fair current should be obligatory. The maintenance of the plants should be given special attention by providing carbonic acid and a good fertilizer regularly in order to obtain healthy growth. An exchange of a quarter to a third of the water every week or two contributes positively to the state of

health of *Pelvicachromis humilis.* If the water has a total and a carbonate hardness of approximately 6°dH and a pH around 6,8, the husbandry preconditions are met. The tank should be illuminated for 12 hours a day throughout the year.

Company may be provided by a pair of a species of *Chromidotilapia* or another *Pelvicachromis*-species and is even recommendable to prevent that aggression is directed towards the *humilis*-partner.

If
breeding

is to happen, the breeding tank should be decorated as described above. Tanks of 70 to 130 cm in length can be used for this purpose. The substrate should consist of a fine dark sort of gravel also in this case. One or two caves are of course required and should be placed directly onto the bottom-plate and be filled up with gravel then. The species digs out its breeding cave by itself and other constructions may collapse by this activity. Using a sub-soil heater is necessary for all substratum- and cave-spawning Cichlids. The heating of the substrate from below causes water to circulate through the gravel and thus avoids a concentration of bacteria which may otherwise later threaten the larvae when they develop in a depression in the substrate. Of course snails are removed from a breeding tank!

In addition to the breeding-caves a couple of other hiding-places must be provided since it may take many weeks until spawning eventually takes place and the female often need to hide from the "amorous" or aggressive male. Dense vegetation is recommended for the same reason. An efficient filter keeps the water clean and rich in oxygen until the larvae swim free. Thereafter it is switched off and replaced by a weaker internal filter to prevent the young fish from being sucked into it by the strong current. The water-quality in the breeding tank should resemble the natural environment as closely as possible. Only then proper development of the eggs and larvae is guaranteed. The temperatures are constantly kept at a level between 26 and 28 °C and the aquarium is illuminated with 0,5 Watt per litre as normal.

One selected pair of *Pelvicachromis* is set into this aquarium. To prevent the fish from becoming shy and hiding in the caves continuously, six to eight Live-bearing Tootcarps or Gouramis should be used to provide company. Since they "inhabit" the upper levels, the available total space is divided equally. The patience of the keeper may however still be stressed, and a is not always breeding attempt successful. The species may start reproducing already at a length of 8 cm. Both partners engage in digging out a cave. Gravel is carried out of it and a shallow shelter with a small entrance providing just enough space for both specimens is created. Pottery flower-pots, with a small opening on the edge and turned upside down, are also suitable and readily accepted as a breedingcave. The female indicates her willingness to spawn by shaking her strongly arched body in front of the male. If the male responds by bobbing up and down and an oblique swimming position with the head down, spawning can be expected soon.

Once the female stops or hardly ever leaves the cave or the flower-pot, one can be almost sure that she will present herself as a mother with a small school of young fish approximately eight days later to guide it through the aquarium jointly with the father. Small amounts of newly hatched nauplii of *Artemia salina* alternating with very fine, powdery flake-food should then be given several times a day. Only a few days later, the small *Pelvicachromis* become more active swimmers and may, guided and guarded by the parents, also explore the higher levels of the aquarium. This is now the right time to introduce a few snails which consume the remains of food accumulated on the ground and thus prevent the quality of water to deteriorate.

The young fish grow relatively slowly despite favourable circumstances and are cared for by the parents for several weeks.

A male of the yet to be described species *Pelvicachromis* sp. from Guinea.

Another colour-morph which resembles *P. humilis* very much, but where too little material is available to determine whether it is a subspecies or just a colour-variety of the afore described species, is presently referred to as

◆ *Pelvicachromis humilis*
Colour-morph "Kasewe"

A different body shape, the colour, the shape of fins, and the smaller adult size are enough characteristics to distinguish between the forms. These fish display a strong transversal banded pattern which extends up onto the fins permanently. The bright light green colours in the areas of the gills and parts of the belly are especially noteworthy. This colouration, which sometimes changes into an iridescent blue is also present in the fins of the females. A pointed appendage in the central part of the caudal fin is a typical feature for the male. The species grows somewhat smaller than the

Distribution of *Pelvicachromis humilis* Colour-morph "Kasewe"

colour-morph "Kenema". Males are fully grown at 11 cm, females reach 10 cm in length.

Pelvicachromis humilis Colour-morph "Kasewe" ♂

The
natural habitat

lies in the Kasewe Forests, in the region of the slopes of the Kasabere Mountains and its buttresses in central Sierra Leone. The fish inhabits small water-courses which cross forested areas. They have a slow current, are very shallow, i.e. only up to 50 cm deep, very clear, rich in oxygen, and narrow, i.e. up to 3 metres wide. In some places, one finds thickets of *Nymphaea lotus,* the Green Tigerlotus, and species of *Anubias,* but generally the watercourses are bare of aquatic vegetation. Submerged branchwork, leaf-litter, and the roots of the trees are the main places for the fish to inhabit. The banks are bordered with emerse plants. Here and there one finds smaller heaps of rocks, consisting mostly of brown and black lavastones. The substrate is composed of dark gravel and fine lava-sand. The water-courses generally lie in the full shade, and only occasionally sunlight reaches the surface. The slightly brownish coloured water is very soft and acidic and very poor in mineral salts.

The colour-morph "Kasewe" does not occur in large numbers in this area. Despite support from many helping hands from the nearby village Moyambawo we only managed to catch five specimens within two days. However, according to Prof. Dr. THYS and Mr. ROLOFF both colourmorphs are found at other localities where they live syntopically. They can be clearly distinguished from all other forms of *Pelvicachromis* by the shape of the head and the body and especially by their colourations. On occasion of a visit to London, I was asked by Dr. TREWAVAS whether I would possibly share the opinion that this species shows more parallels with *Chromidotilapia*

Pelvicachromis humilis Colour-morph "Kasewe" ♀

Table 5

Location:	Kasewe Forest near Moyambawo, about 98 miles southeast of Freetown, Sierra Leone
Clarity:	very clear
Colour:	slightly brownish
pH:	5,9
Total hardness:	below 1 °dH
Carbonate hardness:	below 1 °dH
Conductivity:	5 micro-Siemens at 25 °C
Nitrite:	0,00 mg/l
Depth:	up to 50 cm, usually 10 to 15 cm
Current:	streaming
Temperature:	25 °C
Date:	16.11.1978
Time:	13.00 hrs

than with the genus *Pelvicachromis*. At this point of time I had just returned from Sierra Leone with the newly caught subadult specimens and could say little about this topic. After having kept and carefully studied these fish for two years now, I have to admit that this question was not without reason. *Pelvicachromis humilis* for example chews substrate in search of food the same way as species of *Chromidotilapia* do. The shape of the ventral fins in the female are however a problem. In all species of *Pelvicachromis*, the females always have short rounded or flattened ventral fins whereas *Chromidotilapia* females have long pointed ventral fins the same as the males. The breeding behaviour is another point of distinction. Whilst *Pelvicachromis humilis* is a substratum- and cave-spawning Cichlid with a mother-father-family structure, all species of *Chromidotilapia* known so far are mothbrooders.

In spring 1981 the company Tagis-Aquarium imported another colourmorph where the males displayed a beautiful red to violet colouration in the anal fin and the lower half of the caudal fin. The upper half of the caudal is lemon-yellow, and the rear part of the dorsal fin contained dark transversal stripes. All males share the feature of growing a more or less large lobate appendage from the central part of the caudal fin. It is thus no trait specific to the colour-morph of the Kasewe area.

The untouched harmony of rainforest

► *Pelvicachromis humilis*
Colour-morph "Liberia-Red"

is the prettiest colour-variety of this species and originates from Liberia, the southern neighbour of Sierra Leone. These fish attain a total length of up to 12 cm in the male and differ from other colour-morphs by the con- spicuous deep red coloured fins. This form usually also displays a pattern of transversal bands which begins behind the head and ends on the caudal peduncle. Females are often lacking this banded pattern. They are however brighter coloured in comparison with the pitch black bordering of the dorsal fin being an impressing contrast. Female specimens grow up to 8 cm in total length.

Pelvicachromis humilis Colour-morph "Liberia-Red" ♀

Pelvicachromis humilis Colour-morph "Liberia-Red" ♂

Ever since it was described in 1901 by the British ichthyologist G.A. BOULENGER,

▶ *Pelvicachromis pulcher*
(BOULENGER, 1901)

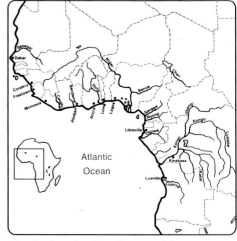

is one of the most popular aquarium-fishes. The bright colours, the interesting behaviour, their small size (males reach 10 cm, females up to 7 cm), and their ability to adapt may be the explanation for this fact. The species has been described under a variety of names and those like

P. aureocephalus
P. camerunensis
P. kribensis
P. subocellatus kribensis

and others still haunt the aquaristic milieu. They were partly based on double descriptions as they were considered different species. In some cases the new names were however base-

Distribution of *Pelvicachromis pulcher*

less. Even scientists were misled, and only the tentative revision of Prof. THYS in 1968 sorted

Pelvicachromis pulcher ♂

130

out the mix-up of names. All these names are today considered synonyms and are no more valid. *"camerunensis"* was previously used for the form *P.* sp. aff. *pulcher* which is described next. *"kribensis"* was applied to the form which is today referred to as *Pelvicachromis taeniatus* Colour-morph "Kienke" originating from the Kienke River system in the vicinity of the town Kribi in southern Cameroon. Although the revision dates back some 30 years, many authors simply do not become used to the correct name of *Pelvicachromis pulcher,* and *"kribensis"* or even *"aquarium-kribensis"* still haunt the literature.

The sexes of this species are easily distinguished. The females clearly grow smaller than the males, have rounded ventral fins, and display a conspicuous broad chrome-yellow band in the dorsal fin.

The species *Pelvicachromis pulcher* includes colour-varieties in which adult specimens generally have no or only a very indistinct lateral stripe on the body or, in contrast, always have a distinct stripe. The black spots, usually bordered golden, are no usable feature for distinguishing between these two forms and may be present in varying numbers. Furthermore, there are other varieties which make this species especially interesting. Those forms where the males display a soft red colouration of the belly or are sound dark olive green coloured are noteworthy. Round or pointed caudal fins, coloured lemon-yellow in one variety, are also of no taxonomic value in this species. One of the prettiest colour-morphs includes males in which the lower half of the body is coloured fire-red from the tip of the snout to the tail. These are however also not to be mistaken for the form which was previously named *"camerunensis"* and is today dealt with as *Pelvicachromis* sp. aff. *pulcher.* All colour-varieties are interesting objects to study.

Pelvicachromis pulcher ♀

The

natural habitat

lies in Nigeria concentrating in the regions west of the Niger Delta around Benin City which is situated approximately 120 kilometres west of the River Niger near the towns of Kwale, Sapelé, and Warri. In the entire area, *Pelvicachromis pulcher* occurs in various colour-morphs in numerous ponds and smaller and larger water-courses.

Two examples should be described in more detail. A small water-course runs south of Benin City. It has a slow current, many pond-like extensions, and is overgrown with mainly emerse plants. The species shares these biotopes with *Hemichromis cristatus, Chromidotilapia guntheri guntheri, Hemichromis fasciatus, Tilapia zilli, Alestes longipinnis,* and *Aphyosemion bivittatum.* The water is clear and water-temperatures range around 27,5 °C. *Pelvicachromis pulcher* obviously prefers the vegetated zones and was observed in large numbers.

The second site is a small river between Benin City and Kwale running south in direction of Sapelé which shall serve as an example of the many clear water-courses. It is the Ethiop River which is a sheer Eldorado for Cichlid-enthusiasts. After a few kilometres from its source it is already up to 15 metres wide and 5 metres deep. It has a quite strong current and its water is crystal clear, without colour, very soft and heavily acidic. The substrate consists of fine laterite-rich gravel which also gives the surrounding land its typical colour. The ground is covered with large groups of aquatic plants of up to a metre in diameter in zones of a more or less strong current with the gaps in between being wide areas of gravel with very few stones, but with partial areas of branchwork or roots. The plant thickets are partly composed of a species of *Vallisneria,* various species of *Nymphaea* including the Green Tigerlotus and the pretty reddish brown *Nymphaea daubenyana* whose off-shoots develop from the leaves. *Pelvicachromis pulcher* was observed living mainly in or on the edge of these plant-islands rarely leaving these sites farther than a metre, returning into its cover

Table 6

A Location:	Branches and pond-like extensions of the small Ogba River, south of Benin City, southern Nigeria
Clarity:	clear
Colour:	none
pH:	4,8
Total hardness:	below 1 °dH
Carbonate hardness:	below 1 °dH
Conductivity:	5 micro-Siemens at 27,5 °C
Nitrite:	not determined
Depth:	up to 60 cm
Current:	very feeble to feeble
Temperature:	27,5 °C
Date:	26.3.1978
Time:	12.30 hrs

immediately at the slightest sign of danger. In between the plants, there were small bare sandy areas which served as territories. The fish had dug small caves between the roots of the plants which were used as hiding-places and most likely also for breeding. In an area of approximately one square metre I counted about 20 subadult to adult *P. pulcher.*

It was observed that females ready for spawning introduced themselves to larger males by arching and shaking their bodies. This obviously prevents the males from venting their natural aggression and driving the female away. It was interesting to note that, when there was no danger around, the *Pelvicachromis pulcher* let themselves drift out of the calm vegetation zone into the slight current calmly waiting for food to come past. I could approach the fish up to a distance of only 10 cm with the hand without any reaction. Only then was the minimum distance reached and the fish dashed back into the covering plant thicket. It

Pelvicachromis pulcher Variety "Green" ♂

Pelvicachromis pulcher Variety "Red" ♂

furthermore appeared that *P. pulcher* and semi-adult *Chromidotilapia guntheri guntheri* of approximately 7 cm in length coexisted in some sort of a food-sharing symbiosis. Whilst the *P. pulcher* picked up food from plants, branchwork, and the ground with the mouth, the *C. g. guntheri* chewed through the substrate for the same purpose.

Other fishes were also observed in the Ethiop River. Occasionally one saw groups of 4 to 6 specimens of *Hemichromis fasciatus* and *H. cristatus* and larger schools of small *Tilapia zilli*. *Tilapia mariae*, as well as *Pelvicachromis taeniatus*, *P.* sp. aff. *subocellatus*, a species of *Procatopus*, *Aphyosemion bivittatum*, small Catfish, and Snakeheads, on the other hand, were not common. The light situation was extreme. At sunlight and under a clear cloudless sky one could measure 60 000 Lux at the water surface and still 20 000 Lux at a depth of 30 cm at 12.00 hrs. Measurements taken at 13.30 with a cloudy sky revealed 25 000 Lux on the surface and 7 000 Lux on the ground in 70 cm deep water. These are light values which are hardly ever achieved in captivity.

Pelvicachromis pulcher does fortunately not belong to those species which are bound to narrow tolerances of water-values. Therefore it is easier than in the case of many other species to adjust the fish to the quality of the water obtained from the tap at home provided it is not too rich in calcium and not too alkaline. A relatively optimal husbandry and the breeding therefore becomes quite easy and many keepers of this species have been surprised by large numbers of offspring thus ensuring this species for the enthusiasts.

Care

When keeping *Pelvicachromis pulcher* certain points should nevertheless be given attention to. A not-too-small aquarium of at least 80 cm in length and 40 cm in depth should be offered since ground-space is very important. The height of the tank is of secondary importance. Fine gravel should be chosen as substrate, and rock constructions and caves are a must. The aquarium should be planted generously

whereby areas in the vicinity of the caves should be left open. Actively swimming fishes which prefer the medium levels of the water and Gouramis for the "top level" are most appropriate for companionship. The often described shyness and furtive behaviour of *P. pulcher* will not be observed under these husbandry conditions.

A powerful filter-system and sub-soil heater are recommendable as the fish are exclusive bottom-dwellers. Those who have soft, slightly acidic water available will easily create an ideal environment. Nevertheless a quarter to a third of the water-volume should be exchanged regularly every fortnight.

The

breeding

requires a well harmonious pair which preferably has shown signs of courtship already in the main tank. They must be transferred into a breeding aquarium which resembles the afore described husbandry environment. Three or four other fishes should be added as companion to prevent the *P. pulcher* from becoming shy. Provided a varying diet of live and flake-food they should soon turn out to be a productive pair. Once "married" the relationship is usually a lasting one, and the preparation of a cave as "wedding-site" announces that an interesting time is to be expected. An ideal spawning-site is once again a thoroughly cleaned hollow coconut-shell with an opening not larger than 3 cm which should therefore be made available and half filled with gravel. An important stage of the preparations for the spawning is that the animals carry substrate out of the cave. The gravel should therefore have a grain-size under 3 mm in diameter since the transport otherwise becomes too difficult for the fish. Furthermore, the small larvae could slip into the gaps later and may even be stuck in there.

During these preparations, the female clearly is the more active partner. She constructs the "kids' room" and additionally "seduces" the male. He himself meanwhile claims his territory and defends it. Eventually

Pelvicachromis pulcher Variety "Yellow" ♂

Pelvicachromis pulcher Variety "Blue" ♂

it comes the time where both of them frequently visit the cave and the courtship in all its bright colours peaks. Thereafter the female hardly leaves the cave any more and one usually sees only her head sticking out of the entrance which is often almost closed by heaps of gravel. The process of development of the eggs hanging on the sidewalls and the ceiling inside the cave is generally concealed from the observer's eyes.

Only after approximately seven days, depending on the temperatures, the female leaves the cave in order to guide a small school of nimble juveniles through the aquarium — jointly with the male which "proudly" presents himself in his brightest colours. This is now the best time for the keeper since nothing is more rewarding than to observe such a fish-family strolling through the tank.

In addition to these already interesting observations, I wanted to study the process of development of the fry in all details. Thus I decided to offer only a stone as spawning-site. I was fortunate enough that the spawning took place on this stone. I photographed it and then gave it back to the parents repeating the procedure in regular intervals to obtain an exact time-schedule. The watertemperature was constantly kept at 29 °C during this time. The female had laid approximately 180 eggs which were immediately fertilized by the male. The spawning act took approximately one hour.

The first eggs (Fig. 1) were taken away three hours later and turned out to be oval in shape. First signs of development were visible. On the slender tip of the egg one could clearly observe the continuous division of the cells. Already two hours later the mono-cellular ovum had split into a clearly recognizable multi-cellular cluster. In order to skip some time in between, the stage 47 hours later shall be described now.

First signs of life were visible. The embryonal fluid passed through some vessels in the egg. Early patterns of vessels had become recognizable on the yolk-sac. In transvision one could clearly make out the embryonic development of a head-like thickening and the embryonic spine. Only another

4 hours later, i.e. 51 hours after spawning, the embryos hatched from the egg shells. The female supported this action by taking the eggs into the mouth or probing them with the lips. The larvae were subsequently transported by the female to a depression inside the cave. There a small heap of larvae piled up which was "lovingly" cared for by an attentive mother. Another 4 hours later (Fig. 2), indications of the head and the main fish-bone had become more distinct. The larvae measured 3,4 mm at this point of time.

After 87 hours (Fig. 3), the larvae had grown to 4,9 mm, and head and tail could be clearly distinguished from the yolk. The latter had clearly meanwhile shrunk. The eyebladder started developing lenses; gill-rifts and gill-arches could be identified. Twenty-four hours later, i.e. 111 hours after the fertilization of the eggs (Fig. 4), the eyes were distinct. The head and the mouth were well developed and the yolk-sac had shrunk even more. Its pattern had become more conspicious. 147 hours after spawning (Fig. 5) the small fish were almost completely developed. One could now recognize the internal organs and that they were fully functional. The head was complete, and even the fins could be identified. A faint camouflaging pigmentation appeared on the body. Only little had remained of the yolk-sac. The larvae, which measured 5 mm, undertook their first swimming attempts.

After 159 hours (Fig. 6), the young fish eventually swam. The camouflaging pigmentation had increased. The swimming bladder and heart were clearly visible. From the spotted pattern of the yolk-sac branched blood-vessels had developed. Measuring the individual young fish showed differences between 5,5 and 6,4 mm. Six days and 15 hours after spawning, at a water-temperature of 29 °C, the development of the young fish was completed.

"Proud parents" took their "children" through the aquarium and nauplii of the Brine Shrimp together with the dust-food MikroMin as first food are the contribution the attentive keeper can make to the well-being of the fish-family.

Egg-development in Pelvicachromis pulcher at a water-temperature of 29°C

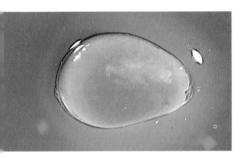

1   Development after 3 hours

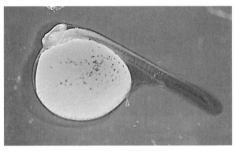

2   Development after 55 hours

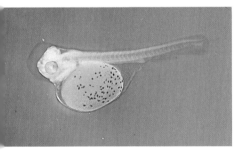

3   Development after 87 hours

4   Development after 111 hours

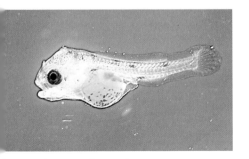

5   Development after 147 hours

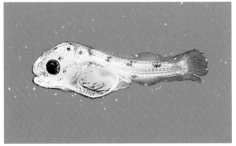

6   Development after 159 hours

Despite controversial discussions, until recently only Nigeria was considered as distribution range of *Pelvicachromis pulcher*. It was not before in the early 1990's that OTTO GARTNER managed to also record *Pelvicachromis pulcher* from Cameroon. His specimens were initially referred to as *Pelvicachromis* sp. aff. *pulcher,* but are today more correctly named *P. pulcher* with the suffix "Ndonga". The natural habitat of these fish lies northwest of Edea, in the area of the settlement Ndonga, and west along the track in the direction of Mouanko, a village near the mouth of the Sanaga. M. FREIER, who confirmed the occurrence of these fish in the described area, made the following data available: pH 4,8 and conductivity 30 microSiemens at a temperature of 25 °Celsius.

O. GARTNER observed and bred these fish in the aquarium and reported about a particular nervousness. The length of adult fishes and their behaviour are not any different from the facts already known, but there are some noteworthy deviations regarding their appearance.

OTTO GARTNER noted the following: "Especially during the reproduction period, both sexes impress, by beautiful colours. The dark band of the body, which is almost permanently visible, extends from the mouth up to the centre of the central rays of the caudal fin. Depending on the disposition of the animals, there may be another stripe of yellow to bronze colour above it which begins above the eye and extends up to the caudal peduncle. The back and upper parts of the body vary between grey and black. A particularity of this form is the fact that males also show a broad bright orange band in the dorsal fin in a state of excitement, similar to that which is characteristic for the females of this species. The dorsal fin of the male is elongate and pointed whereas it is shorter and rounded in the female. Occasionally, males may have a black spot of eye-size in the spinous area of the dorsal fin.

Both sexes have bright yellow bars on the cheeks, lines between the upper lips, the eyes, and in the frontal region. The gular region is light blue, the belly feebly red, and the pectoral

Pelvicachromis pulcher "Ndonga" from Cameroon

fins are uncoloured. The anterior rays of the ventral fins are iridescent blue with the following ones being reddish violet. The anal fin and the lower parts of body immediately above it have a greyish yellow to yellow colouration. The transparent caudal fins of the females are round in shape whereas those of the males are pointed and yellowish orange. The upper section of the caudal fin is spotted dark in the males which I have bred in captivity."

According to HORST MEISTER of Bonoua in Côte d'Ivoire another variety should exist in the region of Bouna in the northeast of this country, i.e. far to the west of Nigeria. These fish were repeatedly caught by H. MEISTER 17 km outside Bouna in direction of Varale, studied, and successfully bred in aquaria in Bonoua. On occasion of a visit to Bonoua we had opportunity to examine these fish. They were *P. pulcher* which are commonly known to us as the Red Variety. No information is available about their natural habitat. Whether the occurrence in Ivory Coast is a natural or artificial one is yet to be determined.

A typical "white-water" biotope in West Africa

Another very interesting but rarely kept species is

▶ *Pelvicachromis roloffi*

(THYS, 1968)

which was discovered by ERHARD ROLOFF. It belongs to the smaller representatives of this genus and is assigned to the *"roloffi-subocellatus*-group". Males attain lengths of 8,5 cm whilst females are fully grown at 6 cm. The distinction of sexes is easy. The males have produced dorsal, anal, and ventral fins. A lateral band of varying intensity runs from the head to the caudal peduncle. Adult females, on the other hand, have a violet coloured belly and almost black rounded ventral fins. Numerous small roundish black spots bordered yellowish golden are usually observed in the vicinity of the bases of dorsal and caudal fins. This spotted pattern is a trait particular to *Pelvicachromis roloffi*.

Distribution of *Pelvicachromis roloffi*

The natural habitat

lies in eastern Guinea, Sierra Leone, and western Liberia. Small pond-like water-bodies and

Pelvicachromis roloffi ♂

partly gently flowing water-courses in plantation areas, bushlands, and forested regions are the biotopes inhabited by this species. In his travelling reports, ROLOFF indicated water-values of 1 to 2 °dH total hardness and a neutral pH ranging around 7. I could only catch this species in the Kasewe Forest on the slopes of the Kasabere Mountains, and the environment was described in detail in the species account for *Pelvicachromis humilis.* Both species occur together here. Notwithstanding this, the recorded waterdata may be repeated here. The water was very soft with a total and carbonate hardness below 1 °dH, the pH ranged around 5,9, and the conductivity was established with 5 micro-Siemens at a temperature of 25 °C. The water was very clean and rich in oxygen, its colour being slightly brownish with a strong torrent at places. *Pelvicachromis roloffi* were observed amongst plants near the banks, submerged branchwork and leaf-litter, and roots exclusively.

Care

One should preferably choose tanks of at least 70 cm, better 130 cm in length. Calcium-free gravel of up to 3 mm grain-size is suitable as substrate. An underfloor heater is recommendable to provide a ground rich in oxygen and poor in bacteria for this species. This is especially important for the development of the fry.

Constructions of calcium-free rocks and thoroughly cleaned coconut-shells with an opening of up to 3 cm in diameter are furthermore required. Flowerpots with an opening on the edge may serve the same purpose. The tank has to be planted generously so that it provides a feeling of safety to the *Pelvicachromis roloffi* and reduces their shyness. A powerful filter should provide clean water rich in oxygen; an exchange of a quarter to a third of the water every week or two is important. The water should be of soft to moderately hard quality and a couple of floating plants, such as the beautiful *Pistia stratiotes,* are appreciated by the fish. It is furthermore recommendable to have some company fishes around like Barbs or Gouramis. A varying diet of mosquito larvae, enchytraeas, flake-food, and daphnia is of importance.

Well cared for specimens in a calm environment inside as well as outside the tank will certainly start

breeding

at one stage. After a lasting period of courtship in which the female especially is very active and presents herself to the male in bright colours arching and shaking the body, spawning takes place. The chosen site is cleared of excessive gravel by both partners and spawning eventually occurs in a "cave". Thereafter the eggs, and subsequently the larvae, are guarded by the female whilst the male defends the area around the cave.

Approximately eight days later the female leaves the cave with a small school of young fish which are henceforth guided and guarded by both parents. At this stage, the young fish are approximately 6 mm in length and have dark pigmentation on the body. If the parents signal danger by jerking the ventral fins, the juveniles remain motionless. Once the danger is over, they continue swimming through the tank under the guidance of the parents, always searching the ground for food. Newly hatched nauplii of the Brine Shrimp *Artemia salina* and fine powder-food should be fed several times a day during the first period.

The powerful filter is replaced by a gentler one to avoid juveniles from being sucked into the machine. The temperature for breeding should be kept at 25 to 26 °C and the water should be of very soft, slightly acidic quality to support a proper development of the eggs and larvae. Three to four days after the juveniles have begun to swim freely, a few snails may be introduced into the tank to consume the remains of food and thus contribute to a healthy quality of the water.

One of the most beautiful small Cichlids of West Africa is certainly

▶ *Pelvicachromis subocellatus*
(GÜNTHER 1871)

which has been kept in aquaria already since the end of the last century. Many attempts have since been undertaken to describe the often changing bright colouration especially in the females. Very impressive examples for this are the drawings by JOHANNES THUMM published in the "Blätter für Aquarien- und Terrarien-kunde" in 1909 and the aquarel painted by CURT BESSINGER in 1942 which shows the female's ordinary, courtship, and spawning colourations magnificently. Even in this time of colour photography they has not lost any of their actuality. Unfortunately, this "old" species is hardly ever offered or kept nowadays.

Two different colour-morphs of a not yet scientifically described species with a similar appearance are often erroneously taken for

Distribution of *Pelvicachromis subocellatus*

P. subocellatus and offered as these. It is a form also referred to as *"Pelvicachromis klugei* II" originating from Nigeria which should correctly be identified as *"Pelvicachromis* sp. aff.

Pelvicachromis subocellatus ♂

subocellatus". The species described by the British ichthyologist GÜNTHER in 1871 has been given a variety of vernacular names. It reaches up to 9 cm in the male and approximately 6 cm in the female. The sexual dimorphism is obvious by adult males having produced dorsal and ventral fins whilst the ventral fins are rounded in the female. Their soft rays of the dorsal fin have a rounded section as well where they usually have a round dark spot. During periods of courtship and spawning, the body of the female becomes almost black with the gill-covers mainly assuming a golden colouration, and a zone of white ranging from the spinous section of the dorsal fin down to the lateral line. There, it changes into a hush of red intensifying to a bright dark red to violet on the belly. When displaying the "normal" colouration, the metallic iridescent yellowish or orange longitudinal band in the dorsal fin is a infallible point of distinction.

The

natural habitat

is indicated by the British ichthyologist G.A. BOULENGER on basis of the type material as Gabon and the rivers Luali, Lundo, and Luculla in the Shiloango region which partially belong to the Angolan enclave between Congo and Zaïre. According to the paper by Prof. THYS VAN DEN AUDENAERDE, the fish are found in the coastal areas of the Gabon capital of Libreville and its vicinity to the south of the country up to Moanda in the lower Congo region. There, they inhabit small current water-courses and ponds with very soft and acidic water.

Care

This species requires a fair-sized aquarium. A dark gravel of fine grain-size is recommendable. Several clusters of rounded calcium-free rocks should be placed directly onto the ground plate and filled up with gravel to enable the animals to create their own caves. Flowerpots and coconut-shells, also partially buried with substrate, may serve the same purpose. A rich vegetation is recommendable as well as some Live-bearing Toothcarps and Gouramis as inhabitants of the middle and upper levels of the tank. The ideal water has a pH around 6 and does not exceed 5 °dH in total and carbonate hardness. The addition of salt, propagated in many publications, is unnecessary and not recommendable. It may be meant to destroy bacteria, but the same goal is achieved more efficiently with a low pH.

The watertemperature should range around 26 °C. For normal keeping, the recommended water-values do not necessarily need to be considered absolute. It is however vital to maintain the water-quality at a high standard by an efficient filtering and sustaining an optimal oxygen-content. A varying diet is of extreme importance.

For

breeding

the temperatures should be increased to 27 or 28 °C and soft acidic water supports healthy development of the fry. After vivacious courtship dances and displaying bright colours, the animals spawn inside a cave. The female assumes an almost black colouration with the gill-covers being yellowish golden and a silverish white saddleblotch appearing above a dark red belly. As is the case in all species of *Pelvicachromis*, it is the female which takes care of the fry and hardly ever leaves it alone. The male meanwhile undertakes to guard and defend the spawning territory. After approximately eight days the juveniles swim free and the parents guide them through the tank. The first food should also in this case consist of newly hatched nauplii of *Artemia salina* and the powder-food MikroMin added by some pulverized TetraOvin. The number of young fish may be between 50 and 60.

After some time of absence from the aquaria, BLEHER managed to again import *Pelvicachromis subocellatus* from the vicinity of the

Pelvicachromis subocellatus Form 1 (Moanda) ♀

town of Moanda in 1986. As was demonstrated by BLEHER, FRECH & MIKSCHOFSKY only one year later, a variety of the Moanda-fish lives in the area of Matadi in which especially the females are different.

Whilst the Moanda-form appears to be somewhat more stout, the specimens from Matadi are more slender and elongate. In order to easily distinguish between these two forms, the fish from Moanda may be referred to herein as Form 1 and those from Matadi as Form 2. Distinct points of differentiation exist especially in the female specimens. Females of the Form 1 display a cherry-red belly with silverish shining scales in its front portion, the entire back and the anal region during court-

ship. The caudal fin is somewhat rounded with little pattern and colour.

In contrast, females of the Form 2 have, besides a black colouration behind the head and on the tail, a feebly reddish or pink coloured belly with the silverish scales being limited to a few ones on the back and the anterior part of the belly. The caudal fin is slightly enlarged in its central part and is thus of oval, slightly lanceolate shape. It is partially distinctly patterned and sometimes bordered black and white.

According to our observations, Form 2 is easier to breed and more productive. The majority of the fish presently kept in aquaria belongs to this form.

Pelvicachromis subocellatus Form 2 (Matadi) ♀

In the first half of 1993, ROLAND NUMRICH of Mimbon Aquarium became the first to import

◆ *Pelvicachromis* sp.
"Guinea"

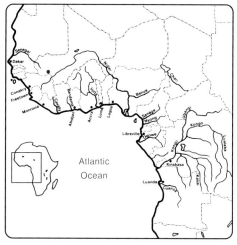

alive. Despite the presumption that this would be a member of the "*roloffi-subocellatus* species-group", all fish imported so far rather show parallels to *Pelvicachromis humilis* of which it might even be a colour-variety. Until this is clarified, these fish should however be referred to as *Pelvicachromis* sp. "Guinea".

The fish dealt with here do not originate from the area of the Bandi River, but from the region of the Kolente River of which part of its course forms the border to Sierra Leone. Their appearance and pattern agrees with the material in the collection of the museum in Tervuren. A specific feature of the females obviously is the obvious large dark blotch on the caudal peduncle. Males, on the other hand, only display black spots on the posterior part of the dorsal fin.

The area of the Kolente River furthermore

Distribution of *Pelvicachromis* sp. "Guinea"

contains a variety of *Pelvicachromis roloffi* and a species of *Hemichromis* which is possibly less aggressive, grows smaller, and has an unusual spot on the body, respectively a dark vertical stripe at mid-body which extends up to the dorsal fin.

Pelvicachromis sp. "Guinea" ♀

145

One of the most beautiful species of the "pulcher-group" is a form which has not yet been scientifically described. As it has no official name and resembles the species *P. pulcher*, it is therefore presently referred to as

▶ *Pelvicachromis* sp. aff. pulcher

It was known for a long time under the trade-names *Pelmatochromis* and *Pelvicachromis camerunensis* respectively after the revision in 1968. The description under the name *P. sacrimontis* which Prof. Thys has been preparing for quite some time, has unfortunately not yet been published. The usage of this name is therefore not yet permitted.

These Cichlids reach the same adult size as the one portrayed before, i. e. 10 cm in the male and 7 cm in the female. They differ distinctly from *P. pulcher*. Their shorter and rounded

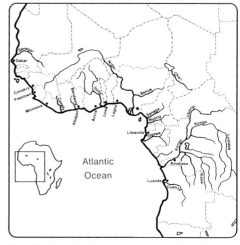

Distribution of *Pelvicachromis* sp. aff. *pulcher*

body-shape together with a bright turquoise striped pattern on the gill-covers below the

Pelvicachromis sp. aff. *pulcher* Form A ♂

eyes are obvious points of distinction, already recognizable in very young specimens of either sex. The males furthermore show a bright deep purple below the dark lateral stripe from the tip of the snout to the anal region or the tail respectively. The remaining parts of the body below this lateral stripe are dark brown to black except for the gill-covers which have a bright turquoise striped pattern on a greyish brown to black ground. The ventral fins are of a dark red colouration with the anterior spines being bright green. Above the longitudinal stripe there is a parallel olive green stripe, and the back is once again coloured greyish brown to black. The females furthermore differ at adult age by having the front half of the dorsal fin coloured blackish grey — a fact which is not found in females of any other species of *Pelvicachromis*. The caudal fin is a feebly yellowish golden in female specimens, carmine-red in the upper half, and bordered with turquoise in

males. Spotted patterns are not known so far, but this possibility cannot be excluded at this stage.

This species also has at least two different colour-varieties of which we consider the one described afore as Form A. Form B is an equally attractive variety and has a beautiful lemon-yellow colouration in addition to the turquoise stripes on the gill-covers, extending up onto the belly where it transforms into a feeble carmine-red. The posterior lower part of the body is a faintly yellowish colour. In both forms the lateral band is predominant at mid-body. Females of the Form B resemble those of Form A, with the colours being not as bright. Furthermore, they show the pretty lemon-yellow as ground-colour on the gill-covers as well as the males of Form B.

The sexes are easily distinguished in both forms. The males grow larger and have pointed dorsal, ventral, and anal fins. The smaller

Pelvicachromis **sp. aff.** *pulcher,* Form A ♀

147

growing females, in contrast, have rounded fins instead. They lack a bright metallic band in the dorsal fin.

The information available about the

natural habitat

of this species is far from being complete. The first specimens of this fish were probably imported by a Hamburg-based pet-shop which also brought up the name "camerunensis". In my opinion, the fish occurs east of the Niger River in Nigeria near the border with Cameroon. Especially the region near the town of Calabar on the Cross River in the northern forest areas may be eligible. It has clear fast flowing water-courses of up to 40 cm in depth with a ground covered by many stones. At certain places the water forces through narrow high banks forming small rapids. Large specimens of the plant *Crinum natans* are commonly found in these streams. The region between Calabar and Uwet, in the direction of Umuahia, is hilly and rich in water. Rainforests and plantations determine the landscape.

Care

One should use tanks with a length of at least 70 cm for this species. The depth of the aquarium should be 40 or 50 cm with a minimum height of 40 cm. Fine gravel of a grain-size under 3 mm serves as substrate also in this case.

An underfloor heater is recommendable in order to keep the ground clean and rich in oxygen. Numerous rock formations, preferably made up of large rounded flat stones, should create caves or be laid flat or obliquely onto supporting stones onto the ground resulting in cavities of 1 or 2 cm in height. They provide the necessary sites for the *Pelvicachromis* sp. aff. *pulcher* to dig their caves which is an important preparatory part of the reproduction!

Two or three pieces of bog-oak should furthermore be used for decoration by placing them onto the stones or arranging them obliquely in the tank and attaching plants like *Anubias nana* or *Microsorum pteropus* onto them. Further vegetation may consist of a large specimen of the Red or the Green Tigerlotus, *Nymphaea lotus*, which are planted in the front or central third of the aquarium. Contrast is provided by using 15 to 20 *Hygrophila stricta*. The Green Tigerlotus stands out best in front of a group of approximately ten *Alternathera sessilis* and a bush of *Myriophyllum matogrossense*.

One should use soft water of 2 to 5 °dH or moderately hard water of 5 to 10 °d carbonate hardness. The pH should range around 6,8 if possible. A strong current is very important for the fish and the plants. Healthy growth of the plants also means that the water is efficiently fortified with oxygen. The aquarium should furthermore be illuminated appropriately by 0,5 Watt per litre of water for 12 hours a day. The recommendable company may be provided by Barbs, Live-bearing Toothcarps, or Gouramis. A pair or two of other species of *Pelvicachromis* or *Nanochromis,* or a pair of a *Chromidotilapia* species, should share their environment. Only then will the *Pelvicachromis* sp. aff. *pulcher* present themselves in bright colours and show their interesting behaviour.

Provided sufficient food in form of live and flake-food and regular exchanges of a quarter to a third of the water every week or two, the otherwise quite shy fish will start

breeding

one day. The patience of the keeper may however be stretched to the limit since it is of utmost importance in this species to have a perfectly harmonizing pair. It is therefore very advisable to keep several specimens together. Once a pair has found each other and shows courtship and imposing behaviour, spawning may take place. A calm environment in front of the aquarium and a balanced territorial behaviour of the fishes sharing the tank have a major impact on this success. A large, preferably oval coconut-shell with an opening of approximately 4 cm in one end and perfectly clean, should be filled to the half with clean fine gravel and then be placed in the aquarium. If this coconut lies in the territory of the courting pair, it is usually readily accepted as "wedding site". The female begins to carry the fine gravel out of the coconut by the mouth which is an activity important for the preparation of releasing spawn. Under vivacious shaking and arching of the bodies both partners then proceed into the cave and stay there for some time. When only the male eventually waits in front of it and the female stays inside with only the head sticking out of an entrance which is almost entirely shut with gravel, one can be sure that spawning has taken place. The male defends the territory in front of the cave. If he can manage this without too many problems, the other fishes may remain in the tank. If the male however has problems with this task, one should remove the "enemy fishes"

three to four days after spawning. Since the female tends to eat the eggs when disturbed, but hardly ever does so with the larvae, this later point of time is obviously more suitable.

If it is tranquil in the breeding tank and the temperature is maintained at 27 °C, the female leaves the cave after approximately eight days as a "proud mother" with a school of young fish. Both parents subsequently guide their offspring very attentively through the aquarium and defend them fiercely when necessary. Newly hatched nauplii of the Brine Shrimp *Artemia salina* and fine powder-food should then be given several times a day.

The powerful filter must now be switched off to prevent the young fish from being sucked in.

Pelvicachromis sp. aff. *pulcher* Form B — top: ♂, bottom: ♀

A similar species which is often mistaken for *P. subocellatus* has been imported more frequently during the past few years. It is yet another species still to be described for which therefore no proper scientific name is available.

▶ *Pelvicachromis* sp. aff. subocellatus

resembles, as this name already indicates, the species *Pelvicachromis subocellatus*. These fish cannot quite compare with the beautiful colours described for the preceding species. In the past, they have been referred to as *Pelmatochromis* and *Pelvicachromis klugei* II of which two colour-morphs are known. The first form grows a little larger with the males attaining approximately 11 cm and the females 7,5 cm in length. Its ground colouration is darker and contains more grey to blue shades. The females never reach the bright colours of yellowish golden and dark red. The second colour-

Distribution of *Pelvicachromis* sp. aff. *subocellatus*

morph grows somewhat smaller and is more similar to the species described precedingly.

Pelvicachromis sp. aff. *subocellatus* ♂, the larger growing colour-variety

The males are fully grown at approximately 9 cm whilst the females do not exceed 6 cm in length. They usually show a feebly red coloured belly and may come close to a *P. subocellatus* female during the display of the bright colours of courtship and spawning, but their colours remain always somewhat more faint and less intense. Also in this case, the females of both forms only rarely have a spotted pattern in the area of the soft rays of the dorsal fin. The sexes are easily determined. As is the case in all *Pelvicachromis* females, there is a bright metallic band in the dorsal fin and the ventral fins are rounded. In contrast, the males have pointed ventral, dorsal, and anal fins.

The

natural habitat

of both forms is apparently not shared. The smaller variety was recorded by myself in the

Table 7

Location:	Oroghodo River on the road from Benin City to Kwale, southern Nigeria
Clarity:	clear
Colour:	slightly brownish
pH:	5,8
Total hardness:	below 1 °dH
Carbonate hardness:	below 1 °dH
Conductivity:	11 micro-Siemens at 25,5 °C
Nitrite:	0,00 mg/l
Depth:	up to 40 cm
Current:	very feeble
Temperature:	25,5 °C
Date:	2.4.1978
Time:	12.00 hrs

Pelvicachromis sp. aff. *subocellatus* (top: ♂, bottom: ♀), the smaller growing colour-variety

Ethiop River in southern Nigeria as described in the species account of *P. pulcher.*

The numerous small water-courses of the area west of the Niger River, in the regions of the towns of Kwale, Sapele, Warri, up to Benin City are the biotopes of this species. As to how far the smaller species possibly prefers the smaller water-courses and the larger one the larger rivers, is only presumptuous.

The description of a biotope on the road from Benin City to Kwale may serve as an example here. It is the Oroghodo River which mouthes into the Ethiop River approximately 30 km farther down. The ground consists of a quartz-sand containing laterite with a grain-size up to 2 mm. It is partially light brown, but mainly dark reddish brown. The water-course partly crosses swamp-like areas and was provided shade by embanking brush and trees. The depth of the water was often only 5 to 8 cm whereas in parts 40 cm were reached. The banks were overgrown by a thicket of plants which mostly also stood in the water. Roots and branches often created an impenetrable submerged jungle. Large specimens of *Nymphaea lotus,* the Green Tigerlotus, had formed large groups with the leaves floating in the current, and a tropical hornwort with delicate leaves had settled in the zones of shallow water. The water was clear, very soft, with a constant slight current, and of slightly brownish colour. Syntopical species of fishes were *Polycentropis abbreviata,* young *Hemichromis fasciatus,* subadult *Chromidotilapia g. guntheri, Epyplatys sexfasciatus, Neolebias ansorgii,* the Red *Neolebias,* and *Procatopus gracilis.* In these shallow waters of approximately only 8 cm in depth, I could also observe large *Hemichromis fasciatus* of some 15 cm in length which stayed motionless in one place waiting for prey to come past. A really interesting biotope.

Care

Aquaria used for this species should not be too small. Moderately sized tanks of 130 cm in length, a depth of 50 cm, and a height of 40 are adequate. A dark fine sort of gravel should serve as substrate. Pieces of bog-oak and coco-nut-shells with an opening of up to 4 cm in diameter should form the decoration together with densely arranged groups of plants leaving open spaces in front of the caves.

A high water-quality is to be maintained by means of a powerful filter and a regular exchange of a quarter to a third of the water-volume every week or fortnight is of extreme importance for the well-being of these fish. One should use only soft to moderately hard water and keep the temperature at 26 °C. Accompanying fishes may be smaller Barbs or Tetras, but also pairs of other *Pelvicachromis*- or *Nanochromis*-species or Gouramis. Otherwise this species may become very shy. Caves and rock constructions have to be numerous to enable all the fishes to inhabit their territories without constant fights.

An aquarium decorated in this manner offers everything necessary for the

breeding

of this species. As is the case with all other species of *Pelvicachromis,* this form spawns inside a cave after an exciting courtship. The fry is taken care of by a mother-father-family and both partners defend their offspring unselfishly. They do not mind attacking even much larger fishes and are usually very successful. For a positive development of numerous young fish, water of a soft, slightly acidic quality is advantageous. On the other hand breedings of smaller numbers have also been successful in moderately hard (5 to 10 °dH) and even in hard, slightly alkaline water of 10 to 15 °dH carbonate hardness. The water-temperature should be increased to 28 °C during this time.

Again, the newly hatched nauplii of the Brine Shrimp *Artemia salina* and powder-food are the ideal diet for the first days. The breeding of this species can be compared with that of the species *P. subocellatus* and *P. pulcher.*

Coming to an end with the portrays of all species of *Pelvicachromis* known so far, I would like to introduce the *taeniatus*-group. Although it contains only one single species, the ten colour-morphs of it make it a fairly large group. Since these colour-morphs inhabit separated areas and differ distinctly regarding colouration and morphology, they might as well be distinguished, in my opinion, at subspecific level. The ten known morphs should be dealt with by using the following adjuncts in order to allow proper determination:

1. "Nigeria-Yellow"   6. "Wouri"
2. "Nigeria-Red"   7. "Dehane"
3. "Nigeria-Green"   8. "Lokoundje"
4. "Moliwe"   9. "Kienke"
5. "Muyuka"   10. "Lobe"

It has subsequently turned out that the subspecies *P. kribensis calliptera* described in 1929 by the French ichthyologist J. PELLEGRIN is yet another colour-morph. According to my investigations, these fish were collected for the first in the year 1926 by THEODOR MONOD of Paris in the area of Dehane in Cameroon. Since the locality name of Grand Batanga was also

mentioned in this connection, it was presumed this taxon may refer to the colour-morph referred to as *Pelvicachromis taeniatus* "Lobe" today. Like all the other colour-morphs it has been collected in the natural habitat by ichthyologists and aquarists during the past years. The names of the colour-varieties derive from the geographical areas in which they occur. For example, the colour-morph "Moliwe" is named after the village Moliwe in western Cameroon, and the form "Kienke" is found in the small branches of the Kienke River which crosses the southern Cameroon port of Kribi. The colour varieties from Nigeria are distinguished by their individual colourations since there is more than one colour-morph of *Pelvicachromis taeniatus* in the country of Nigeria. *Pelvicachromis taeniatus* "Nigeria-Yellow" is the nominate basis of all colour-morphs since the revision in 1968.

An African main-road in the rainy season

◆ *Pelvicachromis taeniatus*

(BOULENGER, 1901)

Colour-morph "Nigeria-Yellow"

This is the form scientifically described as *Pelmatochromis taeniatus* by the British ichthyologist G. A. BOULENGER in 1901. His description was based on material collected by Dr. ANSORGE in 1900 and later added by specimens found by PAUL ARNOLD. In the year 1960, this species was again described as *Pelmatochromis klugei* by DR. HERMANN MEINKEN and re-described as *Pelmatochromis kribensis klugei* in 1965. Both names are however invalid since 1968 and are to be considered synonyms. Even the name *P. kribensis* is invalid and also a synonym of this taxon. This taxon had referred to the colour-morph from the Kienke Riversystem which, though it is distinctly different not only in its colouration, has to be dealt with presently as *P. taeniatus*.

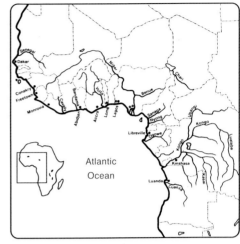

Distribution of *Pelvicachromis taeniatus*

The males of the colour-morph "Nigeria-Yellow" reach up to 8,5 cm and the female 6 cm

Pelvicachromis taeniatus "Nigeria-Yellow" — top: ♀, bottom: ♂

in length. They are distinguished by their remarkably bright and highly contrasting patterns in the dorsal and especially the caudal fins.

Their

natural habitat

lies in the areas west of the Niger Delta in Nigeria. They are said to also occur west of the capital of Lagos in the region between Ado and Awga Awdaw, but it is yet to be determined whether this is really the same colour-morph. It is however fact that they inhabit the Ethiop River and the nearby water-courses. The reference specimens were also caught in this area in 1900. The distribution extends from the towns of Benin City to Kwale, Sapele, and Warri. As to whether this species also occurs on the eastern side of the vast Niger River has unfortunately never been investigated. It is imaginable that, as is the case in other instances, the broad Niger River functions as a geographical barrier preventing this small species from extending its range.

In the Ethiop River, the fish lives sympatrically with several other Cichlid species, e.g. *Chromidotilapia guntheri guntheri, Hemichromis fasciatus, Hemichromis cristatus, Pelvicachromis* sp. aff. *subocellatus,* and especially *Pelvicachromis pulcher,* to mention only some of them. *Pelvicachromis taeniatus* mainly inhabits the shallow zones near the banks. Its localities in the vicinity are also always characterized by low water-levels. The water in this species' biotopes is always soft, very acidic, with a slight current in almost every instance, clear, and generally of slightly brownish colour. Areas with a rich growth of plants near the banks are preferred places to stay. Branchwork and roots as well as the overhanging foliage of emerse vegetation are zones providing the necessary cover and are especially inhabited by small species of fishes. Further details of the natural habitat of *Pelvicachromis taeniatus* Colourmorph "Nigeria-Yellow" may be neglected here as they have been dealt with already in the preceding species accounts of *Pelvicachromis pulcher* and *Pelvicachromis* sp. aff. *subocellatus.*

Care

Keeping this species is not connected to specific problems. Smaller tanks of approximately 70 cm in length are suitable although moderately sized ones of 130 cm by 50 cm in depth and a minimum height of 40 cm are more advantageous for its keeping. The aquarium may be generously decorated and planted since no species of *Pelvicachromis* damages aquatic plants. Only immediately in front of the spawning site some minor "earthmoving activities" may be carried out during which it may occur that small plants could be damaged by piles of gravel. Fine dark gravel with a grain-size of up to 3 mm should be used as substrate.

Rounded calcium-free stones are placed directly onto the bottom plate and arranged to form small caves. The same procedure applies to pieces of bog-oak. Thereafter, the gaps are filled up with gravel covering the underfloor heater. This kind of heating induces a constant heat-flow through the substrate and thus an exchange of water. This is not only advantageous for the plants but also prevents the dangerous development of bacteria in the substrate which may subsequently severely threaten the fry. When a tank is run for a longer time, but even after only a few months, the substrate may contain a nitrite-concentration ten times higher than the "healthy" clean water above it if there is no circulation. Since all species of *Pelvicachromis* are substratum- and cave-spawners, they place their developing larvae in shallow depressions on the ground. There, the embryos are exposed to bacterial influences. A frequent reason why breeding attempts fail although they were successful when the tank was decorated originally, is only due to the fact that there is an insufficient circulation of water through the substrate. The main reason is that the embryos lie directly onto the ground and die from the exposure to a high concentration of Nitrite.

In addition to a decoration with rocks and pieces of bog-oak, the aquarium should be richly planted. Two or three large Green Tigerlotus, *Nymphaea lotus,* should grow in the central areas of the tank. Six to eight stalks of the Sumatran fern *Ceratopteris thalictroides,* four to six stalks of *Limnophila aquatica* with its large leaves, and eight to ten stalks of *Hygrophila polysperma* may be planted in front of the rear wall. *Microsorum pteropus,* the Javanese Fern, can be attached to the pieces of bog-oak with nylon-thread or pinned with glass-needles. The African *Anubias nana* and the Congo Water-fern *Bolbitis heudelotii* can also be alternatingly attached to the rocks. They may also be anchored down onto the ground by small stones. These plants should however not be buried in the substrate since the rhizomes would then die. The number of plants obviously depends on the size of the tank. In any case they require a sufficient supply with nutrients in the form of fertilizers and carbonic acid. A proper filter system inducing a constant current of water is very important. Aeration by means of the common bubble-stone should be avoided since properly growing plants produce enough oxygen. On the other hand it also drives the carbonic acid out of the water. For healthy plantgrowth, an aquarium requires regular and sufficient illumination. The rough formula for this is 0,5 Watt per litre tank-volume for 12 hours daily. An aquarium set up according to the afore mentioned recommendations will suit the requirements of all species of *Pelvicachromis* and the various colour-morphs of *P. taeniatus.* The water-temperature may vary between 24 and 26 °C. One may also connect the thermostat-controlled heater to a timer so that the water cools down naturally by one or two degrees Celsius during the night hours.

For

breeding

soft water-values should be given preference. An aquarium as described for the keeping can be used. The transferred animals will soon adjust to the new tank. If one places several young *P. taeniatus* in it, pairs will form when the specimens reach maturity. Furthermore, if only two specimens of different sex are placed together, the intense courtship behaviour of the female will soon result in a tight and

usually lasting partnership with the available male. It is of advantage to keep a few other fishes with the pair. Another pair of the same or another species of *Pelvicachromis* or *Nanochromis* as well five or six Gouramis or Live-bearing Toothcarps may share this aquarium. Provided with a varying diet in sufficient quantities consisting of mosquito larvae, enchytraeas, daphnia, flake-, and pellet-food, the pair will soon start breeding. Assisted by the male, the female begins to dig a cave and shift gravel from below a rock or in a coconut-shell with only a small opening filled to the half with gravel. After a vivacious courtship during which the female assumes a dark red to feebly light blueish belly, the fish spawn inside the cave. Thereafter the female cares for the eggs and the subsequent larvae leaving the cave only exceptionally. The entrance of the cave is almost closed with gravel and only the head of the female sticks out attentively. Sometimes looking somewhat bored, the male guards the territory in front of the cave. At a water-temperature of 26 °C, the female leaves the cave with a small school of young fish, and both parents subsequently guide their offspring through the aquarium. If no other fishes are kept together in the breeding tank, aggression may have accumulated and can result in quarrels between the parents after a few days. If other fishes share the aquarium, both parental specimens however stand side by side and devotedly defend their territory and their offspring.

The young fish should be fed newly hatched nauplii of the Brine Shrimp *Artemia salina* and powder-food several times a day in the beginning. During the first days the filter-pump should be replaced by a foam-rubber cartridge filter since the wake of the former is dangerous for the young fish. A regular, if necessary even frequent exchange of the water favours the growth of the fry enormously. Notwithstanding this, no more than a third of the water-volume should be exchanged at a time since it may cause drastic and thus dangerous changes in the water-quality.

The reproduction of all colour-morphs of *P. taeniatus* follows the described procedure and only the water-quality determines the number of descendants. Therefore, the water-values of the natural habitats should be emulated as closely as possible. As can be noted from the descriptions of the natural biotopes, three to five colour-morphs of *P. taeniatus* live in very soft and acidic water. Since most of the specimens available to us are either wild-caught ones or their immediate descendants, the tolerance of the eggs and larvae towards the water-quality is still very narrow. This may be outlined by an example.

My attempts to breed *P. taeniatus* Colour-morph "Lobe" revealed that approximately 20 young fish may be expected per clutch in moderately hard to hard water with a carbonate hardness around 12°dH and a pH around 7,8. In other attempts the same pair however produced approximately 80 juveniles in soft to moderately hard water of 5°dH carbonate and 8°dH total hardness, whose pH was reduced to 6,2 by filtering over peat.

Another side-effect was especially note-worthy and also applies to other species of *Pelvicachromis*. Those breeding experiments with colour-morphs of *P. taeniatus* and other species of *Pelvicachromis* which inhabit soft, strongly acidic waters in nature and were now exposed to slightly alkaline water with a pH around 7,8 resulted in the majority of offspring being female. Only 5 to 10% of the young fish were males. Experiments in strongly acidic water with a pH around 5,3 had the opposite result of only 5 to 10% being female.

Discussing this subject with the local fish-collectors of that time during my travels in southern Cameroon, they unanimously confirmed that they were not able to export the ordered *P. taeniatus* of the colour-morphs "Lobe" and "Kienke" occurring there at equal sex ratios due to the simple fact that they would constantly catch fewer females than males. In the respective region the pH varies between 4,8 and 6,5. My experiments furthermore revealed that a pH of 6,2 results in an almost balanced ratio of sexes.

The topic of sex-determination by changes in the concentration of hydrogen-ions is, in my opinion, a rarely investigated subject — at least there has been written little about it. It is however a fact that the concentration of

hydrogen-ions, the pH in other words, influences the determination of sex. Every aquarist should therefore take these experiences into consideration when breeding a specific species and attempt to breed males and females at a balanced ratio. This positively contributes to the conservation of the species for our hobby and their further promotion.

Clear and uncoloured water-courses are an infrequent appearance in West Africa.

▶ *Pelvicachromis taeniatus*

Colour-morph "Nigeria-Red"

By sorting wild-caught specimens from Nigeria and partly also from old breeding stocks, the "Red Ones" became known. These specimens undoubtedly belonged to the species *P. taeniatus* thus there were already three different varieties from Nigeria. Firstly there was the well-known yellow colour-morph, in which male specimens had yellow coloured gill-covers and chests. Secondly there was the green morph whose gill- and chest-region was coloured green with a varying intensity but which entirely lacked any yellow shades. And thirdly, there was this red colour-variety whose gill- and chest-zones, but especially the gill-covers and mostly even the entire lower head was coloured a deep blood-red.

The female specimens of all three colour-forms also show interesting traits of distinction. Whilst the yellow and, according to J. FREYHOF, the green form inhabit numerous streams and small rivers in the area between the towns of Benin City and Onitsha, no con-

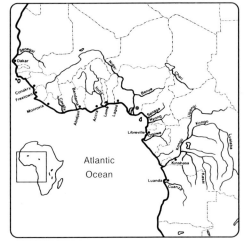

Probable Distribution of *Pelvicachromis taeniatus* "Nigeria-Red"

firmed locality data is available for the red morph. According to information received from an importer, the green variety also occurs in the vicinity of Calabar in the frontier area to Cameroon. In my opinion this is however

Pelvicachromis taeniatus "Nigeria-Red" — top: ♀, bottom: ♂

doubtful. One can do nothing but guess where the red colour-morph might occur naturally. Where might they really live?

So far, several different forms of *Pelvicachromis* have been discovered which have an extreme red colouration. Firstly, there is a very bright red coloured form of *Pelvicachromis pulcher* whose males are coloured intensely blood-red from the tip of the snout to the tail. They also originate from Nigeria, but their precise range is as yet also unknown. Another Dwarf-cichlid with an interesting red colouration is *Pelvicachromis* sp. aff. *pulcher* "Form A" which has not yet been described scientifically and been referred to as *Pelvicachromis camerunensis* for a long time. Already in 1966, KLAUS KLUGE reported about the catching of these beautiful fish in the coastal areas around Calabar in the extreme southeast of Nigeria. My own studies in the region between Onitsha, Port Harcour, and Calabar in 1982 could however not reconfirm the occurrence of this species there. I could not even record specimens of this form in the deep brownish coloured waters of the swamp and forest

areas along the road from Calabar to Mampfe (Cameroon). Relying on the reports of KLAUS KLUGE it would be possible that the "Red Ones" occur in this area. Unfortunately we have to be satisfied with this hypothesis. Only the local collectors know the precise localities — and they keep their knowledge for themselves.

According to the present state of knowledge, *P. taeniatus* Colourmorph "Nigeria-Red" is an easy-to-keep fish which breeds readily in captivity. Whilst moderately hard to slightly alkaline water is however fully sufficient for the keeping, soft and slightly acidic water is required for its breeding. This red colour-variety is comparatively productive with just under 100 young fish per clutch being the average. According to experiences made so far, almost 100 percent of the juveniles become as brightly red as their parents. Adult males attain lengths of approximately 8 cm whilst females are fully grown at some 6 cm.

This red colour-morph is an appreciated enrichment of the unfortunately still very few species of Cichlids available from West Africa.

Pelvicachromis taeniatus "Nigeria-Red" ♂

▶ *Pelvicachromis taeniatus*

Colour-morph "Nigeria-Green"

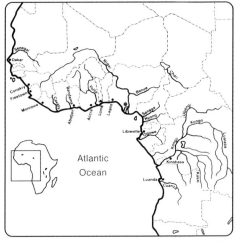

Since the late 1980's another colour-variety of *P. taeniatus* is known from Nigeria, commonly referred to as "Nigeria-Green". These are specimens whose gill- and chest-areas are neither yellow nor red but exclusively green. The females show a red blotch on the belly with the upper region being greenish blue. Both sexes furthermore have black spots bordered light in the soft rays of the dorsal fin and in the upper half of the caudal fin. This colour-variety resembles all other forms from Nigeria with regard to size, husbandry, and reproduction. Following unconfirmed reports by importers, this form originates from the area of Calabar. Despite this, Jörg Freyhof was able to record it from the region of Warri, west of the Niger, in the northern delta of this river. This is the area where the type specimens were also collected originally.

Distribution of *Pelvicachromis taeniatus* "Nigeria-Green"

The colour-morph "Nigeria-Green" is quite productive and its breeding has been successful on numerous occasions. It is nevertheless still rarely seen in aquaria.

Pelvicachromis taeniatus "Nigeria-Green" — top: ♀, bottom ♂

◗ *Pelvicachromis taeniatus*

Colour-morph "Moliwe"

resembles the varieties "Nigeria" to a great extent, but is nevertheless distinctly different. The approximately 8 cm long males are comparatively smaller and have more red in the dorsal fin, but hardly any dark round spots in the soft-ray section. The caudal fin lacks the dark bordering in the lower part, and the anal fin is patterned with light spots instead of the light blue and reddish brown colouration. The females attain lengths of 5,5 cm and are thus distinctly smaller. They have fewer dark round spots in the soft-ray sections of the dorsal and caudal fins with the latter lacking the yellowish striped pattern. During courtship the belly assumes a dark red, violet, or feebly

Distribution of *P. taeniatus* "Moliwe"

Pelvicachromis taeniatus "Moliwe" ♂

161

brilliant blue colouration. The ventral fins are rounded or flattened respectively. The bright longitudinal band of chrome-yellow colouration in the dorsal fin is an easily recognizable feature to distinguish sexes.

The
natural habitat

lies in western Cameroon. This colour-morph is probably endemic to a water-course up to 4 metres in width and in parts up to 80 cm in depth which crosses the road from Victoria to Kuma between the villages of Moliwe and Mile 4. The stream cuts through plantations and secondary forests. This colour-variety is very common here and was caught by various col-lectors within a period of several years in this water-course only. It shares its habitat with *Chromidotilapia finleyi* Colour-morph "Moliwe" which is probably endemic to this stream as well. The water is clean and very clear. Its values are 4 °dH total and 3 °dH carbonate hardness, pH 7,6, conductivity 130 micro-Siemens established at a temperature of 26 °C.

The few water-courses of this area are generally separated from each other by geographical barriers, i.e. by the buttresses of the Cameroon Mountains. This is a possible explanation for the differences in the individual populations of one and the same species within a rather small area. The biotopes of the forms portrayed in the following are situated similarly close to each other.

Pelvicachromis taeniatus "Moliwe" ♀

◆ *Pelvicachromis taeniatus*
Colour-morph "Muyuka"

This colour-variety is of the same size as the colour-morphs of Nigeria yet the pattern and colouration are entirely different. The males reach up to 8,5 cm in length and have a more contrasting green pattern on the gill-covers and only a few, often very indistinct spots in the upper part of the caudal fin. The females are fully grown at 6 cm and display a reddish to violet blue belly alongside with a single dark spot in the upper part of the caudal fin during courtship. As is the case in all colour-morphs of *P. taeniatus*, their ventral fins are rounded or flattened respectively. The bright chrome-yellow longitudinal band in their dorsal fin is another point of sexual distinction.

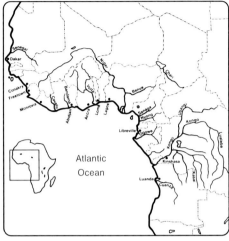

Distribution of *Pelvicachromis taeniatus* "Muyuka"

The

natural habitat

of this form also lies in western Cameroon. It is probably endemic to a water-course only 20 km north of the habitat of *P. taeniatus* Colour-morph "Moliwe" crossing the road from Victoria to Kumba between the villages of Muyuka and Malende. This stream oozes away after only a few kilometres and mouthes into the Mungo River subterrestrially. The water-values resemble those indicated for the preceding form.

Pelvicachromis taeniatus "Muyuka" — top: ♀, bottom: ♂

◆ *Pelvicachromis taeniatus*

Colour-morph "Wouri"

The local form of *P. taeniatus* that turned out to be the most difficult to keep was the one originating from the Wouri area. Additionally these fish appear to especially susceptible to a variety of diseases in captivity. The few successful breedings could not contribute to a wider promotion of these fish in aquaria. OTTO GARTNER and his friends were the first to record this form in the area of the Nkwoh River near Wouri northeast of Duala. The study site is accessible from the road to Yabassi, approximately 17 km east of the turn-off from the road connecting Duala with Edea. There, they managed to catch several specimens of *P. taeniatus* of the local variety "Wouri" amongst swamp and aquatic plants and the following

Distribution of *Pelvicachromis taeniatus* "Wouri"

Two courting ♀♀ of *Pelvicachromis taeniatus* "Wouri"

water-values were established: pH 6,0 and 30 micro-Siemens at a water-temperature of 26 °Celsius.

On several occasions, the colour-morph "Wouri" has been described as being especially beautiful and interesting, but this is possibly only due to its rarity. But is this form really so beautiful? OTTO GARTNER described its appearance as follows: "Male *P. taeniatus* of the form "Wouri" display a narrow white margin with a successive broad black zone consisting of four to six fused spots on the hind edge of the upper half of the caudal fin and another white bordering to the base of the fin. The lower half of the caudal fin is bordered with a narrow dark band followed by a row of blueish spots on red ground on the inner side. The other fins, except for the glassy transparent pectorals, are feebly coloured. The body of the fish is grey, becoming lighter towards the belly, and the scales are bordered dark.

Female specimens of the form "Wouri" display themselves in colourfulness which can hardly compared with anything seen before! At the height of excitement the body assumes a blueish black colour. The mid-body is multicoloured, i.e. orange in the upper zones, then iridescent green, and eventually dark red on the belly area. The dorsal and caudal fins are yellowish orange with a black bordering. A small black caudal spot bordered light − its distance to the base of the fin is smaller than to the end of the fin − is probably the typical feature for the females of this form. Yellow lips, greenish yellow gill-covers, and dark red ventral fins complete the splendid appearance of "Wouri-females". We have nothing to add to this.

Pelvicachromis taeniatus "Wouri" ♂

◗ *Pelvicachromis taeniatus*
Colour-morph "Dehane"

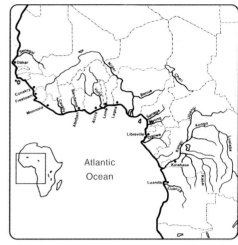

Distribution of *P. taeniatus* "Dehane"

In the year 1990, possibly for the first time, ZEISS and FREYHOF managed to import a fish alive which had already been described by PELLEGRIN as *P. kribensis calliptera* in 1929 based on material collected by T. MONOD of Paris in the region of Dehane in Cameroon in 1926. Since then, these fish were referred to as *Pelvicachromis taeniatus* form "Dehane". JÖRG FREYHOF published the following on this subject: "The locality lies approximately 1 km off Dehane, a village on the lower Nyong. It is a small stream which crosses the road. On the 26.2.1990, the water had hardly any current, the biotope was shaded, and the ground was covered with leaf-litter. *Pelvicachromis taeniatus* was quite common here and we caught specimens of all ages. Only very large and very small ones were missing." The area of Dehane is situated between Edea and Kribi.

With regard to colouration and pattern, the form "Dehane" resembles the variety "Kienke". Males show a bright red colouration in the fins and conspicuous black spots in the upper half of the caudal fin. In contrast to the form "Kienke", the males of the variety "Dehane" additionally show a red colouration of varying intensity on the gillcovers. Females usually lack the spots on the fins, but have a brilliant blue belly with a red tinge.

The colour-morph "Dehane" is not without problems regarding breeding in captivity.

Pelvicachromis taeniatus "Dehane" ♂

◗ *Pelvicachromis taeniatus*

Colour-morph "Lokoundje"

One of the most spectacular discoveries of the past few years is certainly the local form "Lokundje" of *Pelvicachromis taeniatus* which was found by OTTO GARTNER and his friends in the river-system of the Lokundje. Unfortunately however the majority of specimens available for studies was female. Their colouration is very attractive with the belly being coloured red bordered light blue during periods of courtship. The upper parts are iridescent light. The soft-rayd section of the dorsal fin and the entire caudal fin lack any pattern and colour. They do not display any spots or blotches whilst males have a conspicuous pattern and colouration. OTTO GARTNER noted the following immediately after catching: "The upper lobe of the caudal fin contains two rows of three black spots each bordered light (in one specimen only two each), each of which having approximately the size of the eye. The lower

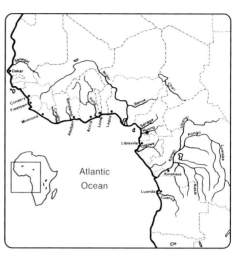

Distribution of *P. taeniatus* "Lokoundje"

half of the caudal fin, the dorsal, and the anal fins have some shade of red patterned with transversal rows of light blue spots."

Pelvicachromis taeniatus "Lokoundje" ♀

167

According to statements made by the collectors who recorded these fish for the first time in 1989, the

natural habitat

lies northeast of the port of Kribi in the south of Cameroon. OTTO GARTNER noted the following: "On the last day of our stay in Kribi, we proceeded for 40 km along the coast to the north turning right at the village of Fifinda. From there, a little used laterite-track leads towards Song Mbong. The beautiful area along the road is hilly and still mainly vegetated with primary forest. After some time we reached a fork-junction, and taking the right extension we reached the first stream after approximately 1 km hill-down. It runs from north to south through Mbihé mouthing into the Lokundje.

On the site, ERWIN KUBER established the following values: Water-temperature 24° Celsius, conductivity 15 micro-Siemens, pH 5,7."

Pelvicachromis taeniatus ♂ caught in the area of the Lokoundje River in Cameroon by R. SAWATZKI in January 1993.

Pelvicachromis taeniatus
Colour-morph "Kienke"

s the form BOULENGER described as *Pelma-ochromis kribensis* in 1911, the name derived from the village of Kribi and the Kribi River, which, for the past 70 years, is known as the Kienke River. Its widely branched system crosses the village of Kribi in southern Cameroon and mouthes into the Atlantic Ocean.

The males of this morph usually attain a length of 8 cm, but exceptionally may even grow up to 9 cm. The have mainly orange coloured fins, occasionally furnished with rows of small light spots, and usually have three roundish dark spots on the upper part of the caudal fin which is bordered fire-red. The females reach 5,5 cm in length and lack the dark round

Distribution of *P. taeniatus* "Kienke"

Pelvicachromis taeniatus "Kienke" ♂

169

spots in the caudal fin, but have a spot in the hind part of the dorsal fin instead. For a long time this feature was considered the only point to distinguish them from females of the colour-morph "Lobe" (compare the following account). During courtship, the belly turns agate-grey to brilliant blue. Their ventral fins are rounded or flattened respectively. The bright longitudinal band of chrome-yellow colour is a good trait to distinguish between the sexes.

The

natural habitat

of this colour-variety lies in southern Cameroon. It lives in the small northern and south-ern branches of the Kienke River, in the vicinity of the village of Bandevouri in direction of Lolodorf in the north, and in the region of Angalé, Akok, and Akom II in direction of Ebolowa. These water-courses are exclusively of small size reaching up to approximately 5 metres in width with a maximum depth of 80 cm, but are usually only 20 to 30 cm deep. They run through the shade of rainforests. The ground consists of fine gravel of ochre to reddish brown gravel. Patches of 3 to 4 metres in diameter are vegetated here and there by *Nymphaea lotus*, the Green Tigerlotus. Shallow zones of only 10 cm in depth near the banks or in nooks of the water-courses with a dense vegetation and submerged leaf-litter and branchwork are favourite places for the fish to

Pelvicachromis taeniatus "Kienke" ♀

stay. These small areas are usually more exposed to light since they are bare of bushes and trees. The shaded zones on the other hand rarely show any aquatic vegetation. Occasionally, small groups of *Anubias nana, Anubias lanceolata,* and *Bolbitis heudelotii* are found. Since the streams meander through the rainforest most of the way and fallen trees block the current here and there, many underwashed sites have formed on the banks. This has also caused the roots of trees to be cleared of the soil and now form an almost impenetrable labyrinth. All these places are favourite sites for the *P. taeniatus* to live. The water always has a slight current which in some places even become torrential. It is always very clear and clean with the intensity of the brownish colour varying in places.

During my studies I could establish that a pair of *P. taeniatus* — one often finds them in pairs — inhabited an area of 20 square metres occupying a territory of 4 square metres. Only juveniles and subadult specimens are tolerated by the dominant pair. Flying prey, i.e. insects, but mainly water-bugs, water-spiders, and the juveniles of other species of fishes form the diet in addition to the large numbers of crayfish inhabiting these waters. Almost daily rainfalls, even in the so-called dry season from November to April, keep the water fresh and clean. The values established in these areas permit a comparison to distilled water. Two examples: the first site is on a water-course with a swamp-like extension and emerse vegetation in the northern Kienke area near the village of Bandevouri. The total as well as the carbonate hardness was below 1 °dH, the pH measured around 6,0, and the conductivity was established to be 20 micro-Siemens at 27 °C. The slightly higher water-temperature was due to the more sun-exposed situation of the swamp-like area. The measurements were taken at 14.30 hrs in the afternoon.

The second example is a water-course with an equally very red laterite-rich substrate near Akok. Here I also could establish a total and carbonate hardness below 1 °dH, a pH near 5,8, and a conductivity of approximately 10 micro-Siemens at 24,5 °C. The measurements were taken at 10.30 hrs in the forenoon.

This watercourse cut through the forest and was rarely exposed to direct sunlight.

All wild-caught specimens of *P. taeniatus* colour-morph "Kienke" could only be adjusted to moderately hard water with limited success. Breeding was much more successful in soft water. The hybridization with other colour-morphs, especially the forms "Lobe" and "Nigeria" were successful under captive conditions. One should therefore ensure that only specimens of the same colour-variety are kept together for breeding in order to guarantee the purity of the forms.

Large specimens of the Water-lily *Crinum natans* grow in current waters.

171

▶ Pelvicachromis taeniatus

"Nange"

Unfortunately some colour-varieties have become an asset of aquarists and are considered colour- or local forms whose status is doubtful. The specimens furnished with the adjunct "Nange" for example are hybrids between the known colour-morphs "Lobe" and "Kienke" which occur naturally. The term "Nange" refers to the local main-stream systems in the natural distribution range. With a distance of only approximately 6 km in the mouthing area, they are very close to one another. The fish do however not live in the large main rivers but almost exclusively in their smaller branches and small tributaries. The southern tributaries to the Kienke and the northern ones to the Lobe River are often separated by just a few metres in their upper regions. Cutting through slightly hilly areas,

Distribution of *Pelvicachromis taeniatus* "Nange"

they sometimes even run parallel to each other near their mouthing zones. This becomes very

Pelvicachromis taeniatus "Nange" ♂

obvious at many places when one travels the rainforest road from Kribi eastward to Ebolowa. A small water-course called Nange is amongst the southern tributaries to the Kienke. Since this region is rich in water anyway, even slight manipulations, e.g. by the local people, may result in a connection of the water-courses and thus in an opportunity of different population of a species to mix.

This is the geographical situation. But, how does this colour-variety differ from the forms "Lobe" and "Kienke" known so far? Since these fish were imported for the first time by OLE SEEHAUSEN, numerous breedings showed that varying ratios of "Lobe"-like and "Kienke"-like specimens were produced. The former are recognized by usually lacking the spotted pattern in the dorsal and caudal fins in both sexes and the latter having larger black spots bordered yellow and red in the upper part of the caudal fin in the males and black spots in the soft ray-section of the dorsal fin in

the females. There are however also intergrades between these colour-varieties. Occasionally it was said that the "Nange"-males could be distinguished by having a green outer bordering of the lower part of the caudal fin. Although this feature is distinctly visible in most "Nange"-specimens, it is not characteristic since males of the morphs "Lobe" and "Kienke" also have this green bordering though in varying intensities.

Summarizing the aforesaid one may state that this is no new colour-variety, but that one should further apply the adjunct "Nange" to these animals. Only by applying this can it be ensured that the pure "Lobe", "Kienke", and the intergrade "Nanga" are kept separate although it is actually one and the same species.

Biotope of *Pelvicachromis taeniatus* in the area of the Kienke-Lobe River-systems in southern Cameroon

▶ *Pelvicachromis taeniatus*

Colour-morph "Lobe-Red"

Surprisingly enough, this colour-morph, frequently kept and bred over many generations, has changed its appearance as it is known to date. Is this a result of cross-breeding or of ongoing inbreeding? It is in any case a fact that the yellowish golden colouration of the throat and chest of the males of *Pelvicachromis taeniatus* "Lobe" has been successively replaced by a greyish yellow and more and more deep red colours have appeared on the gill-covers. Even dorsal, caudal, and anal fins have become deep red to wine-red today.

Our studies revealed that it is impossible to separate both these colour-shades since one and the same clutch contains the original golden yellowish variety as well as the new "Lobe-Red". Only by means of a strict systematic selective breeding could one stabilize the new, very attractive colour-variety. Is it inbreeding or possibly the result of a hybridization with the local form *P. taeniatus*

Distribution of *P. taeniatus* "Lobe-Red"

"Dehane"? It would be a pity if *P. taeniatus* was mixed by inconsiderate or deliberate cross-breedings. On the other hand — *P. taeniatus* "Lobe-Red" is an attractive colour-spot in an aquarium.

Pelvicachromis taeniatus "Lobe-Red" ♂

▶ *Pelvicachromis taeniatus*

Colour-morph "Lobe"

This chromatic aberration is the most conspicuously distinct one of all forms. The males do not have any dark round spots in either the dorsal or the caudal fin. Male specimens reach up to 8 cm in length and show an intense shade of golden yellow on the gills, the chest, and the belly which changes into the colour of honey or a shade of brownish beige towards the back. The caudal fin is marked with a fire-red stripe bordered pearl-white along the upper edge. The females reach 5,5 cm in length and display a agate-grey to brilliant blue belly during courtship. Their ventral fins are rounded or flattened respectively. A bright longitudinal band of chrome-yellow colour ornaments the dorsal fin and is a sure trait to distinguish the sexes.

The natural habitat

of this species also lies in southern Cameroon. The numerous small and narrow branches of

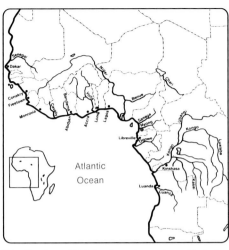

Distribution of *Pelvicachromis taeniatus* "Lobe"

the Lobe which mouthes into the Atlantic Ocean approximately 7 km south of the port of Kribi, resemble those described for the morph "Kienke" and may even rarely be identical. Since the southern tributaries of the

Pelvicachromis taeniatus "Lobe" ♂

175

Kienke and the northern watercourses of the Lobe are often separated by not more than a few metres in the some 7 km wide mouthing delta, locally restricted floods caused by heavy rainfalls or human influence may connect watercourses and thus enable one colour-variety to intrude into the habitat of the other. In the southern tributaries, the fish live in even softer water than described before. Near Eboundja, between Kribi and Campo, we established a total and a carbonate hardness below 1 °dH, a pH of 4,8, and a conductivity of only 10 micro-Siemens — which even decreased to 6 in more remote areas — at a water-temperature of 24,5 °C. These biotopes are shared by the species *Chromidotilapia finleyi* Colour-morph "Campo".

With the portrayal of the variety "Lobe", the chapter on Pelvicachromis taeniatus may be closed now. It is however pretty sure that not all existing chromatic aberrations have yet been discovered. Intergrades from localities such as "Kumba-Funge" and "Nanga" still require more intense and long-term studies before they might possibly one day be considered local varieties. "Lokundje" and "Dehane" on the other hand show many parallels and resemble each other so closely that they might even belong to the same colour-morph. At present, we thus know 10 clearly distinguishable local or colour-varieties of *Pelvicachromis taeniatus*, i.e.

Colour-morph "Nigeria-Yellow"
Colour-morph "Nigeria-Red"
Colour-morph "Nigeria-Green"
Colour-morph "Moliwe"
Colour-morph "Muyuka"
Colour-morph "Wouri"
Colour-morph "Dehane"
Colour-morph "Lokoundje"
Colour-morph "Kienke"
Colour-morph "Lobe"

Pelvicachromis taeniatus "Lobe" ♀

The

Genus Steatocranus

presently unites eight species which differ from all other West African Cichlids by their appearance and their style of swimming. They inhabit torrential water-courses and often even raging sections of rivers. This includes rapids and zones of wild water which cut through gulleys between or over rock barriers. Since they have evolved over the time to true bottom-dwelling fish, their swimming bladders are somewhat reduced thus preventing them from swimming freely in the water and making them unlike most other members of the family. Supported by the ventral fins, they usually "sit" on the ground.

The name *Steatocranus* refers to the overall image of these fish. All species have a conspicuous reservoir of fat on the forehead with the head looking oversized and the mouth being large and high-set. This feature is referred to in their collective common name lumpheads of lionheads. At the first glance they may look like predatory fish, but they are quite tolerant towards other, even small fishes and often look somewhat "sad" and "bashful".

Nevertheless they may sometimes become quite agitated amongst each other.

Five of the eight species originate from the lower parts of the Congo River. Just one species really inhabits West Africa and is thus separated from the other members of the genus by several thousands of kilometres. A few species have been kept in the tanks of enthusiastic aquarists whilst the others are probably known from preserved material only:

1. *Steatocranus casuarius*
2. *Steatocranus gibbiceps*
3. *Steatocranus glaber*
4. *Steatocranus irvinei*
5. *Steatocranus mpozoensis*
6. *Steatocranus rouxi*
7. *Steatocranus tinanti*
8. *Steatocranus ubanguiensis*

Three of these species were a part of the results of a scientific expedition and described as new by the American scientists ROBERTS and STEWART in 1976. As far as is known so far, all species belong to the cave-spawning Cichlids which care for and rear their descendants in a mother-father-family structure. After spawning the females are clearly the more active part in these species.

"Travelling aquarists" — often a support of science

The most familiar species is

◗ *Steatocranus casuarius*
POLL, 1939

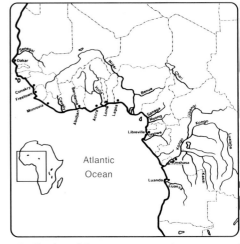

Distribution of *Steatocranus casuarius*

where especially the males built up a prominent bulge on the forehead as age increases. Due to the conspicuous mouth with its sturdy lips they have an almost "threatening" look. They reach approximately 10 cm in nature, but may grow a little larger in captivity. The females have a less developed "lumphead" and grow a little smaller. A accurate determination of sex is only possible once specimens are large and almost fully grown. Then, the males have attained the larger bulge, are bigger, and possess longer dorsal and anal fins. In contrast, the females attain a wider girth; they were

Steatocranus casuarius ♂

repeatedly mistaken for a different species and named *Steatocranus elegans*. This name is however faulty in every aspect probably being based on a confusion with *Steatocranus elongatus* which is a synonym of *Steatocranus rouxi*.

The
natural habitat

ies in the region of the lower Congo River and ts tributaries in Zaïre. According to most recent studies, specimens of these fish were recorded from all over the lower parts of the Congo River from the village of Kinsuka south of Kinshasa up to the village of Inga some 280 km farther down lying approximately 50 km off the town of Matadi in the tip of the funnel of the vast Congo Delta. Here, the fish live amongst rocky areas in the calm water-zones of the rapids and wildwaters of the main river. Crevices and caves in and between the rocks are preferred sites for the fish to live. The water-temperatures in the main river vary between 25 and 29 °C between the end of June and early September. The water is very rich in oxygen; 8,0 mg O^2/l were established. The pH varies between 7 and 7,5.

Care

This species requires appropriately sized aquaria. Tanks of a minimum length of 130 cm and a depth of 50 cm may be deemed adequate. The substrate should consist of fine gravel with a grain-size of up to 3 mm. Several stones, placed directly onto the bottom plate, should be made available as hiding-places. Earthenware tubes with an inner diameter of approximately 50 mm, neatly arranged between rocks, have also proven very useful since they are readily accepted as hiding-places and spawning sites. For reasons of safety these tubes should however be open from both sides and not be blocked by other objects. A generous and dense vegetation is also highly recommendable since it provides a feeling of safety to these often very shy fish. The results are a better development of the fish and enhanced opportunities for observation.

The water should not be too hard and very rich in oxygen. At a temperature around 26 °C, the water should have a slight constantly current, and always be clean and clear. Regular exchanges of a quarter to a third of the total volume of the tank at intervals of a week or fortnight improves the water-quality. Other fishes may be kept with them, and only in cases where other bottom-dwelling Cichlids compete with them for a too small ground-space complications may arise. Provided with a harmonizing pair,

breeding

should occur at some stage. This is preceded by excessive "excavation works" by both parents during which the chosen spawning cave is dug out. It is therefore crucial to ensure that these caves, especially in breeding tanks, contain enough gravel which can be transported.

After a splendid and vivacious courtship, the animals spawn on the ceiling of the cave. The eggs are relatively large and are cared for by the female. The exact time required for the development of the fry has never been determined since the young fish leave the cave only a few days after they have begun to swim free. Thereafter, the parents guide them through the aquarium. Both parents and the male especially protect the descendants even after for several weeks.

Their first food should consist of the nauplii of the Brine Shrimp *Artemia salina*. Since the young fish also pluck off leaves and the ground for food, one should also supply powdered flake-food regularly. During this time the quality of the water is a crucial factor since it is of outstanding importance for healthy growth of the juveniles. If provided a proper environment, the school may contain 80 to 100 young fish which may already measure approximately 2 cm after four weeks.

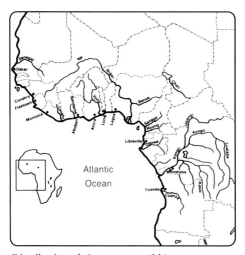

Distribution of *Steatocranus gibbiceps*

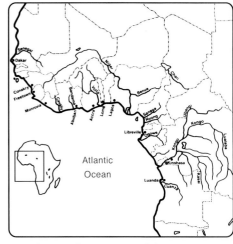

Distribution of *Steatocranus glaber*

▶ Steatocranus gibbiceps

BOULENGER, 1899

has probably never been imported alive. The species has a similar body-shape and apparently has the same body-height as *S. casuarius*. This species has a slightly smaller bulged forehead and a somewhat larger but narrower snout with two very enlarged canine teeth in the upper and lower jaws. These fish are coloured olive brown and, in contrast to *S. casuarius*, the scales are light in their centres and bordered dark. The gill-folds are uniformly dark brown or blackish brown. The chest is a creamy colour. This species presumably reaches approximately 10 cm in length.

The

natural habitat

is situated in the area of the lower Congo River between the villages of Kinsuka and Inga in Zaïre. The water-temperatures here vary between 25 and 29 °C during the course of a year. The oxygen-content ranges around 8,0 mg O^2/l and the pH varies between 7,0 and 7,5. The fish is supposed to inhabit zones of calm water between rocks or in the mouthing areas of small tributaries.

▶ Steatocranus glaber

ROBERTS & STEWART, 1976

has been described only quite recently and is probably known only from preserved material. This is the reason why no description of the colour in life is available.

The few preserved specimens appear light brownish and greenish. The centres of the scales are light and are bordered dark resembling the species *S. gibbiceps*, but being less contrasting. In life, these fish probably have feebly reddish fins, whilst gill-folds and chest may be light brownish or dark creamy.

The body-shape appears to be somewhat stouter by the body being shorter, the caudal peduncle more slender, and the caudal fin longer. The species is of same height as *S. casuarius* and *S. gibbiceps*. The maximum size is not known, but should range around 10 cm in length.

The

natural habitat

lies in the vicinity of the village of Inga on the lower Cameroon River and its tributaries. It resembles that of the species portrayed preceedingly.

▶ *Steatocranus irvinei*

(Trewavas, 1943)

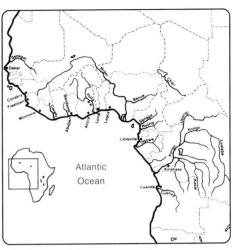

The body is coloured dark greyish green intensifying on the head and the gill-folds. The dorsal and anal fins are dark brown transforming into blackish on the chest and the belly. This species has an overhanging snout and a shortened lower jaw indicating parallels to the species *S. mpozoensis*. This trait is otherwise not found in any other species of *Steatocranus* in the region of Zaïre. These fish grow fairly large and may reach 15 cm in total length in their natural biotopes. Scales are visible on the gill-covers and the head is comparatively narrow with the eyes being big. Another point of distinction are the 11 to 12 gill-rakers on the first gill-arch of which the upper two or three are soft.

Distribution of *Steatocranus irvinei*

The

natural habitat

of this species lies in West Africa in the stricter sense, several thousand kilometres northwest of the distribution of all other species of *Steatocranus*. The reference specimens were collected in the area of the Senchi ferry on the river Volta in Ghana. Another six specimens were caught in the same river in the area of the dam of Akosombo.

Steatocranus irvinei

◆ *Steatocranus mpozoensis*

ROBERTS & STEWART, 1976

Live specimens of this recently described species have a greenish coloured body and reddish fins. The centres of the scales are light with the edges being dark. The gill-folds are shaded yellowish green.

In relation to the upper jaw, the lower jaw is shorter so that the upper teeth are free and even visible from below when the mouth is shut. The species furthermore differs from *S. casuarius* and *S. ubanguiensis* by the relatively shorter fins and a different colouration. They furthermore differ from *S. gibbiceps* and *S. glaber* by having an outer row of smaller and divided teeth. Compared with *S. irvinei*, these fish have fewer gill-rakers and larger eyes.

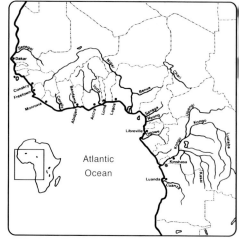

Distribution of *Steatocranus mpozoensis*

The

natural habitat

lies in Zaïre. The reference specimens were collected in the rapids and the torrential waters of the Mpozo River. The exact locality is indicated as the area 5 km down the river from the bridge over the Mpozo, on the road from Matadi to Kinshasa. Only a few kilometres east of the town of Matadi, the Mpozo mouthes into the Congo River.

Steatocranus mpozoensis

182

◆ *Steatocranus rouxi*

(PELLEGRIN, 1928)

has also not yet been imported alive. A few features can be used to distinguish it from other species. For example, these fish differ from all other species of the genus occurring in the lower Congo River by the number of scales around the caudal peduncle; while these generally count 15 to 16, this form has only 12. *Steatocranus rouxi* is very slender and thus resembles *S. tinanti* though the body is somewhat higher. The scales are light in their centres bordered dark. Regarding this trait they resemble the species *S. gibbiceps, S. glaber,* and *S. mpozoensis*. The species *S. rouxi* was sometimes also referred to as *S. elongatus* NICHOLS & LA MONTE, 1934. This name is however considered a synonym today.

The

natural habitat

was indicated to be Luluabourg-Kasai in Zaïre. These are watercourses of and tributaries to

Distribution of *Steatocranus rouxi*

the Lulua River in the region of the town of Luluabourg in the Kasai District, but not the Kasai River which, as a tributary to the Congo mouthing near the village of Kwamouth, completes its long course partly as frontier river between the countries of Angola and Zaïre.

Steatocranus sp. — a remarkably slender species

◆ Steatocranus ubanguiensis

ROBERTS & STEWART, 1976

has rarely been imported alive. Nevertheless, a few features are known to determine it. The fish has an extremely steep forehead, a shorter mouth, less teeth, and fewer scales on the front part of the body and the entire belly. As is the case in *S. casuarius*, the scales are dark with light edges, thus differing from all other species of *Steatocranus*. Although similar to *S. casuarius*, *S. ubanguiensis* is distinguished by having shorter fins, a shorter caudal peduncle, a shorter head, and smaller eyes; it does however have the same number of scales along the lateral line.

In preserved specimens, the body is dark brown with the head also being dark brown or blackish. The gill-folds are brownish in the centres with a distinctly set-off thin black edge. The maximum-length is unknown.

Distribution of *Steatocranus ubanguiensis*

The natural habitat

lies near Gozobangui in the Mbomon River, a tributary to the Ubangui, the frontier river between Congo and Zaïre, and subsequently between Zaïre and the Central African Republic.

Steatocranus ubanguiensis

Steatocranus tinanti
(POLL, 1939)

is a very slender species which has repeatedly been imported into Europe alive, but only in small numbers. It is entirely different from *S. casuarius* by the forehead bulge being very small and the mouth very large. The determination of sex is very easy in adult specimens. Males have a bigger head and produced dorsal and anal fins. They may reach up to 13 cm in length; the females are somewhat smaller.

The

natural habitat

was indicated as the region of Leopoldville, the Kinshasa of today, capital of Zaïre. This refers to the Congo River at the exit of the Stanley Pool, today known as Pool Malebo.

Distribution of *Steatocranus tinanti*

Steatocranus tinanti ♂

185

Care

Tanks with a spacious ground-surface should be used for this species. Aquaria with measurements from 130 cm in length and 50 cm in depth ontoward are the minimum requirement. The height of the tank is of secondary interest, but should not fall short of 30 cm. The species appreciates many hiding-places. Therefore, many rock-piles should be placed directly onto the bottom plate which, filled up with fine washed gravel, provide a large number of caves and hiding-places. A generous vegetation is also advantageous since it provides further cover.

The water should be as soft as possible and neutral in its pH. The water-temperature may range around 26 °C. As the fish inhabit water-courses with a strong current in nature, a strong movement of the water in the aquarium is also recommendable. It should always be clear and clean and especially rich in oxygen. Powerful food, such as mosquito larvae, small earthworms, and enchytraeas, has to be the main part of the diet schedule.

The

breeding

can cause problems. The harmony between the partner is of extreme importance. Once a pair has bonded, it is usually the female that moves into the cave of the male. After some "modifications", during which mainly gravel is shifted from here to there, often lasting several days, the specimens spawn inside the cave. The relatively large eggs are cared for by the female, constantly fanning the clutch in order to provide fresh water. An older female may lay up to 100 eggs. For healthy development, the oxygen-content of the water as well as the hardness, nitrite-, and pH-values are critical. At a temperature of 26 °C, the embryos hatch after five days and swim free another seven days later. They spend another week inside the cave and only then leave it — guided and protected by both parents. At dusk, the school of

juveniles returns to the cave. Also in this case, the first food should the nauplii of *Artemia salina*.

The species of the

Genus Teleogramma

are the Cichlids of the real rapids of the Congo River-system. Four species are presently recognized to belong to this genus. They hardly differ from each other and only one is known to have been kept in captivity. It is however not sure whether the one or the other species has been unknowingly imported amongst the small numbers of *T. brichardi* which sometimes reach the pet-shops. This genus is endemic to Zaïre and exclusively contains Cichlids of elongate, quite flat shape with small roundish scales. The longest soft ray in the ventral fin is either the third or the central one. Generally, these fish grow up to 12 cm in length, with the females being somewhat smaller. The narrow red edge on the peaks of the dorsal fin is characteristic. It extends over the soft-ray area up to the upper edge of the caudal fin. Particular to the males is that the caudal fin is usually uniformly dark whereby a narrow light band may be present on the upper edge. These features are not known from any other species of Cichlid.

One of the four species of *Teleogramma* originates from the Kasai River, all the others inhabit the rapids of the congo River in Zaïre.

The four species presently recognized to belong to the genus *Teleogramma* are:

1. *Teleogramma brichardi*
2. *Teleogramma depressum*
3. *Teleogramma gracile*
4. *Teleogramma monogramma*

The best-known, or rather probably the only one kept in aquaria is

◆ *Teleogramma brichardi*
POLL, 1959.

These slender, usually greyish brown to graphite-black coloured animals have a high degree of intraspecific aggression. They therefore require a large area of ground for their territories. Other bottom-dwelling species are tolerated better than specimens of the same species. Released aggression does however not necessarily mean that serious bites resulting in injuries occur. One should nevertheless ensure that every specimen has an area of 50×50 cm available. Male specimens attain a length of up to 12 cm; females are somewhat smaller.

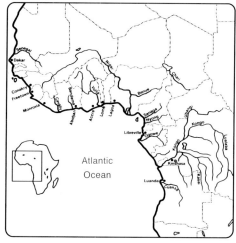

Distribution of *Teleogramma brichardi*

Teleogramma brichardi ♂

187

Determining the sexes is relatively easy. In addition to the difference in adult size, the females have a broader white upper bordering of the dorsal and the caudal fin especially. An additional red edge, as mentioned in the generic description, is not recognizable in this particular species. These fish are poor swimmers; they usually "sit" on the ground supported by their ventral fins. The colouration changes with the mood. In addition to a sound black, they often assume a mouse-grey or brownish grey colour with dark transversal bands. They inhabit caves and constantly monitor their territories. If elevated objects, such as rocks or pieces of wood, are accessible in the immediate vicinity, which allow the fish to "sit" on, these are used as "viewpoints". If other fishes of the same species intrude into the territory or food is spotted, the otherwise poor swimmers can dart through the water in a flash. *T. brichardi* is a "lone wolf" which only accepts partnership during the reproduction period.

The

natural habitat

is the rapids and zones of wild water in the lower Congo River-system in Zaïre and especially the region around the village of Kinsuka southeast of the city of Kinshasa below the large Lake Malebo. Here, water-temperatures of 28 to 29 °C were established. The water is very rich in oxygen, has a low degree of hardness, and a pH between 7 and 7,5.

Teleogramma brichardi ♀

Distribution of *Teleogramma depressum*

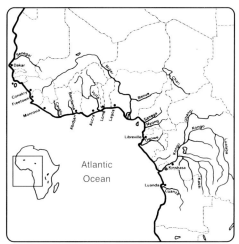

Distribution of *Teleogramma gracile*

Unfortunately very little is known about the other species of this genus.

◗ Teleogramma depressum
ROBERTS & STEWART, 1976

differs distinctly from its allies by the body-shape. It has a very flat head, and the body-width is approximately three and a half times the body-height. The spines and rays of the dorsal fin are clearly shorter than in all other *Teleogramma*-species. Females display a very fine black line on the upper edge of the caudal fin. This species has the highest number of scales along the lateral line.

The

natural habitat

of these fish is the rapid area near the village of Inga on the Congo River in Zaïre. By reports on other fishes one knows that the Congo River, occasionally also referred to as Zaïre River, has water with approximately 8,0 mg Oxygen per litre and a pH between 7 and 7,5. The water-temperature is on average 27 °Celsius.

◗ Teleogramma gracile
BOULENGER, 1899

is the type species of the genus. It is very slender with the spines and rays in the dorsal fin being somewhat longer than in related species. On the other hand, it has fewer rays in the anal fin.

Only in this species the females have a broad black edge above the red zone in the upper part of the caudal fin. This trait conspicuously distinguishes them from *T. brichardi* and *T. depressum*. The maximum length of adult specimens is unknown.

The

natural habitat

of these fish are the torrential waters of the lower Congo River near the village of Bulu, approximately 180 km southeast of Kinshasa in Zaïre. Large boulders lie in the water and break the torrent of the Congo River which thus offers water rich in oxygen and calm zones on the ground and between the rocks as suitable biotopes for this species. They share their habitats with species of *Nanochromis* and *Steatocranus*. Data on water-values are indicated in the descriptions of other species of this genus.

Distribution of *Teleogramma monogramma*

▶ **Teleogramma monogramma**
(NICHOLS & LA MONTE, 1934)

is the fourth species and almost unknown. It differs primarily in morphological aspects. Besides a differen colour pattern, morphometric data serve to distinguish it. The colouration of the upper part of the caudal fin in the female is unknown. The species has not yet been imported alive.

The

natural habitat

are areas of the Lulua River near the town of Luluabourg in the Kasai District in Zaïre.

The species of the

Genus Thysochromis,

belong to the somewhat larger growing Cichlids which are occasionally also referred to as Five-spotted Cichlids because of their four to five dark lateral spots. This common name is however unfortunately also used for other fishes and confusion with *Hemichromis fasciatus* and *Hemichromis elongatus* as well as *Tilapia mariae* is often the result. All three mentioned species differ with regard to body-

shape and especially in the larger total length. The generic name *Thysochromis* was introduced by DAGET in 1988 to substitute the preoccupied name *Thysia*. Both names were to honour the Belgian ichthyologist and director of the Royal Museum for Central Africa in Tervuren, Dirk THYS VAN DEN AUDENAERDE. The genus contains species which were previously assigned to *Pelmatochromis*, i.e.

> *Thysochromis annectens*
> *Thysochromis ansorgii*
> *Thysochromis arnoldi*
> *Thysochromis maculifer*

The paper of DAGET titled *"Thysochromis* nom. nov. en remplacement de *Thysia"* was published in the periodical Mutanda Ichthylogica in 1988.

Pistia stratiotes is a commonly found plant in Africa

◗ *Thysochromis ansorgii*
(BOULENGER, 1901)

◗ *Thysochromis annectens*
(BOULENGER, 1913)

have been considered identical by several authors since the indicated points of distinction were considered too minor. In comparison *T. ansorgii* however has a less pointed head than *T. annectens,* five or more instead of only four blotches on the flanks and never has a blueish green ground-colour. The old names of *Pelmatochromis arnoldi* and *Pelmatochromis ansorgii* are invalid today and considered synonyms of *T. ansorgii. Thysochromis* annectens replaces *Pelmatochromis annectens* and *Pelmatochromis maculifer.* The latter mentioned synonym is probably based on a duplicate description by Dr. ERNST AHL of the Zoologisches Museum Berlin in 1929 who had one specimen available in alcohol.

Distribution of *Thysochromis ansorgii* and *T. annectens*

Determining the sexes of sexually mature specimens is easy. Males attain a length of

Thysochromis ansorgii ♂

191

approximately 12 cm whilst females are some-what smaller and fully grown at some 10 cm in total length. The latter have a red belly of vary-ing intensity which is flanked in the anal area by some bright silverish scales. This is a trait which can also be found in females of *Nano-chromis dimidiatus*. So far it was repeatedly presumed that these bright silverish scales are a usable feature to determine females. Unfortu-nately, this feature is also present in males although the brightness is considerably less intense. Males have a light spotted pattern in the area of the soft rakes of the dorsal fin as well as in the caudal and anal fins which is mis-sing in females. Furthermore, females have slightly rounded back sections of the dorsal and anal fins which are pointed in males. According to the original description, the

natural habitat

of both these species of *Thysochromis* lie in the coastal areas west of the Niger Delta in southern Nigeria. The Niger River itself or smaller tributaries in the west are usually indi-cated as collecting sites. AHL referred to a com-munication that the species described by him should have probably originated from Lagos. Collecting and study reports from other West African countries however record these species also from the territory of Côte d'Ivoire, i. e. the Ivory Coast, up to the eastern forest regions of Ghana. The bordering countries of Togo and the Republic of Benin are also known to host these species, thus the distribution is closed along the entire West African coast. Which of the species lives where has not been deter-mined so far. They are unfortunately usually treated as one and the same.

Care

Keeping both species is not difficult. Aquaria of 130 cm in length and 40 cm in depth are ade-

Thysochromis ansorgii ♀

quate. Fine gravel of up to a grain-size of 3 mm should be used as substrate. Large rocks have to be placed here and there forming vertical or oblique surfaces. Pieces of bog-oak are further means for the decoration of the aquarium. Larger groups of plants should be set up. The observation that the species of *Thysochromis* would damage or eat plants cannot be confirmed by myself. It might however happen that plants are dug out when the spawning territory is created or that they are damaged otherwise, but this cannot be interpreted as intolerance towards plants. The fish can often be observed plucking food-particles from plants and rocks and vegetarian components should form part of their diet.

Both *Thysochromis*-species are known as timid Cichlids. Both reach maturity at a length of approximately 6 cm. When keeping a group of young specimens, one may observe that pairs are formed. This is also the time when territories are claimed and ferocious quarrels may occur. Only if a sufficiently sized aquarium allows several pairs to set up their territories, it is unnecessary to remove excessive specimens. Accompanying fishes are strongly recommendable. Representatives of other families, but also Cichlids of the genera *Pelvicachromis, Nanochromis,* and *Chromidotilapia* may be kept with them to prevent the *Thysochromis* from becoming shy. Regular exchanges of a quarter to a third of the water-volume every week or fortnight have a positive effect on the health of these species as well. Powerful food added by vegetarian components should make out the diet.

The

breeding

of *Thysochromis*-species is interesting, but hardly differs from the strategies of other West African Cichlids. The breeding tank should not be too small, i. e. aquaria of at least 70 cm in length and 40 cm in depth should be chosen. The readiness of a pair to spawn will soon be observed by their claiming a territory and preparations for the spawning act. Oblique or vertical rock surfaces, overhanging rocks or wood, but also roof-like arrangements and caves are preferred places for spawning. If necessary, even the glass-surfaces of the aquarium are used. The "couple" cleans everything thoroughly with the mouth. Whilst the female especially focuses on the spawning site, the male patrols the territory. In order to stimulate the caring and protection instinct of the animals and to prevent an accumulation of aggression which may be caused by the lack of potential predators on the fry, a few "co-inhabitants" should be kept in the breeding tank.

These species spawn while displaying their finest colours. Although being generally considered substratum spawners, they actually tend toward a cave-spawning strategy. They often accept shelters where they, as most of the West African Cichlids do, attach their eggs to the ceiling. The eggs may number up to 500 in adult specimens. In many cases, the patience of the keeper is excessively tested, since the clutch is often eaten after a short time. This is rather normal in young specimens which spawn for the first time. Should it however be repeated again and again, it is recommendable to verify the watervalues. An overpopulation of the breeding tank might be another possible reason, i. e. the parents do not have the necessary tranquillity to care for the fry. Experiments revealed that soft and slightly acidic watervalues increase the number of juveniles. A water-temperature of 25 °C is recommendable. Whilst the female devotedly guards the clutch, always fanning fresh water around it causing the eggs to waver on their small stalks in the current, the male digs some shallow depressions in the ground. He however spends most of his time with defending their territory. At this stage he is however clearly the more passive partner. After approximately three days, the embryos break through the egg-shells. The female takes the larvae into her mouth and transports them to one of the depressions in the substrate. This may also be any depression in front of a stone or between plants. During the course of the following days, the fry is relocated repeatedly. After approximately seven days, the young fish begin to swim. Now, also the male also becomes more active. As a "proud father" he joins in with the mother to guide the school of young fish which initially stay hesitantly close to the ground, but later rise to higher levels. Run-aways are caught with the mouth by both parents immediately and brought back to their siblings. At dusk, the juveniles are returned to a depression in the ground where they spend the night protected by the parents.

Infusoria and fine powder food are recommended as first food since the juveniles are initially very small. A subtly bubbling aeration should cause a slight current in order to keep the food-items moving. The powerful filter-system which otherwise initiates a constant current should be switched off or its inlet be covered in a way that it does not threaten the only 4 mm long juveniles. Nauplii of the Brine Shrimp *Artemia salina* may be offered after three to four days. Only then will one recognize the enormous appetite of the tiny fish. After approximately four weeks they may have grown up to 2,5 cm in length. A good water-quality is essential for healthy growth and repeated partial exchanges of the water therefore strongly recommendable.

Once the parents are observed to engage in new preparations for spawning, the juveniles should be removed. The species of *Thysochromis* do unfortunately not belong to the colourful fishes with a strong demand and a wide acceptance by the aquarists. Nevertheless they are an interesting alternative for the advanced keeper.

If one would try to portray the large

Genus Tilapia

in its whole, it would certainly blast the margins set by this book. The species of this genus which was established in 1840 furthermore grow large, too large for a husbandry in an aquarium. Exceeding 20 cm in length in several instances or due to a specific behaviour they are usually kept only by specialists. Their general distribution in the tanks of the hobbyists is very limited.

On the other hand their behaviour is usually very interesting. Unfortunately they often lack the "manners" necessary if the keepers likes a well decorated aquarium with aquatic plants. The species of *Tilapia* are sub-stratum-spawners. They usually spawn on a firm ground, but sometimes also directly on the fine substrate of a large pit. The offspring in cared for in a parental family. In large parts of Africa several species play an important role as food for man. They have also been introduced in many parts of Asia as protein-supply and to enhance the general sustenance situation. That these "useful fishes" affect the natural fish-fauna in many areas and alter it is an unpleasant side-effect for the ichthyology.

In order to at least give a superficial impression of this genus, three species are dealt with here. These are relatively unknown Cichlids which are however appropriate for husbandry in comparably small aquaria and which may attract the interest of keepers due to the interesting behaviour and appearance.

Wanted Large-scaled Tetras for the pot

One of these is

Tilapia busumana
(GÜNTHER, 1902)

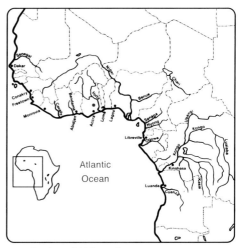

Distribution of *Tilapia busumana*

whose appearance may be considered that of an "average" *Tilapia*. A characteristic feature for this species is the dark blotch in the posterior part of the dorsal fin. Otherwise its appearance is typical for the genus. Specimens of the populations inhabiting small watercourses in rainforests attain a total length of approximately only 10 cm whereas those from the populations of the Volta Dams, other lake-like water-bodies, and the large rivers reach 20 cm in length. The sexes differ regarding body-shape and colouration. Adult males have a more prominent forehead and have lost the typical blotch in the dorsal fin. Their ground-colouration remains somewhat lighter than in female specimens. During periods of courtship, the females show a faint dark broad longitudinal and transversal striped pattern on the body and a sound blueish green ground colour.

Tilapia busumana

This species is generally tolerant towards other fishes and is considered "plant-friendly". Nevertheless, minor damage can be caused especially to young sprouts since these fish also require vegetarian components in their diet. An additional supply with vegetarian flakefood, such as TetraPhyll, is therefore crucial.

The
natural habitat

lies in water-courses cutting through the rain-forests and plantations of the vast Kumasi region in southern Ghana. These are usually small, up to 3 m wide, and in places up to 1 m deep streams. Specimens from here only reach up to 10 cm in length. The larger growing specimens of the same species inhabit the rivers of Bia, Tano, Ofin, Birim, and Pra as well as the Volta Lakes, larger lake-like water-bodies, and the isolated Lake Bosumtwi. Certain authors have questioned the identity of the fish of the Lake Bosumtwi described in the travel-report of J. PAULO in 1976 and figured by P. LOISELLE at another place so that it may be doubted that this really is the same species. Specimens from Lake Bosumtwi are more blue in colour, but otherwise have the same body-shape. The specimen figured here originates from the Birim River and was undoubtedly determined as *T. busumana* by Prof. Dr. THYS VAN DEN AUDENAERDE. It is clearly a substratum-spawner. Adult specimens of the river-populations of approximately 10 cm in length lay some 400 eggs per reproduction act.

Stages in the Development of eggs in Tilapia busumana

The photographs indicate that this is a substratum-spawner.

The eggs shortly after they have been laid. The clutch contains appr. 400 eggs.

After appr. 24 hours at a temperature of 26 °C the first dark spots appear.

Only 5 hours later, i.e. 29 hours after spawning, the first larvae break through the egg-shells.

Approximately 4 days later, being 125 hours old now, the development of the embryos is completed and young fish swim free.

▶ *Tilapia buttikoferi*
(HUBRECHT, 1881)

is a larger growing species reaching a total length of approximately 25 cm. It has a conspicuous and attractive pattern consisting of deep dark brown to black transversal bands on the body with the interspaces being yellowish in young animals and silverish grey in adult specimens. This species is not always good-natured and especially larger specimens may be a danger to smaller fishes. They like to make usage of shelters under rocks or wood as places to stay. Only hardy plants such as species of *Anubias* and *Bolbitis* should be chosen since these fish often feed on plant material. They furthermore rigorously remove and bite off softer plants when establishing a territory during courtship and spawning season. The spawning takes place in *Tilapia*-manner and the parents care for their fry in a parental family-structure. Powerful food should be obligatory and contain earthworms, mosquito-larvae, fish-roe, enchytraeas, and vegetarian components in short intervals, e.g. TetraPhyll.

In the year 1968, Prof. THYS described a new and similar species under the name of

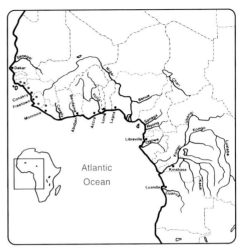

Distribution of *Tilapia buttikoferi*

Tilapia cessiana which partly occurs syntopically, but is somewhat smaller in adult size.

The

natural habitat

are water-courses between Guinea-Bissau to western Liberia.

Tilapia buttikoferi

197

◆ *Tilapia joka*

THYS, 1969

is the *T. moorii* for the friend of West African Cichlids. The original description depicts this Cichlid as a *Tilapia* with a "round head and a small mouth". It is unfortunate that these fish have been imported for the enthusiasts so rarely. The species is adequate for keeping in a well-decorated aquarium with a rich growth of plants and may be socialized with other West African Cichlids without causing problems. In my specimens I could not yet manage to determine a sexual dimorphism. They appear to be fully grown at 10 cm in total length. Their rounded head gives them a good-natured appearance. Their behaviour is peaceful. With increasing age, the strongly banded pattern temporarily changes into a blueish grey with a conspicuous scale-pattern. Although one can depict them as plant-tolerant fish, vegetarian components should not be missing from their diet. Attention should be paid to a permanently clean water rich in oxygen in order to keep this species healthy. For their breeding, soft acidic water is presumably a precondition.

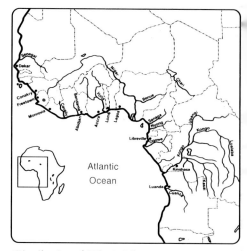

Distribution of *Tilapia joka*

The

natural habitat

of *Tilapia joka* lies in Sierra Leone. The fish inhabit water-courses in the region of the villages of Pujehun, Gobaru, and Bayama in the south of the country.

Tilapia joka

PHOTO CREDITS

Photographs: Horst LINKE

except for:

Otto Gartner: page 138, 164, 167

Jürgen KNÜPPEL: page 45, 63

Roland Numrich: page 58

Hans-Joachim RICHTER: page 84, 85

Erwin SCHRAML: page 188

Wolfgang Staeck: page 10, 11, 182, 184

Drawings on pages 81, 118, 119
Koniglik Museum van Mittel-Afrika, Prof. Dr. Dirk Thys van den Audenaerde

Reference Index in Volume II

THE AUTHORS

HORST LINKE, born in 1938, has had an interest in the aquarium since early childhood. Already quite early the dream of all enthusiastic aquarists to visit the tropical habitats of our aquarium-fishes came true for him. In the year 1963 he undertook a journey throughout Black Africa, and two years later he had opportunity to visit the countries of Panama, Venezuela, Peru, and Bolivia. From the contacts with the aquarium-fishes in the wild, new questions and tasks always developed so that he visited some countries not only once but repeatedly. Beginning in 1973, he undertook collecting and study expeditions to Cameroon, Nigeria, Ghana, Togo, Sierra Leone, Tanzania, Kenya, Thailand, Sumatra, Borneo, Malaysia, Colombia, Peru, and Bolivia in quick succession. During his numerous stays abroad it was always of special interest for him to collect as much information as possible about the life-conditions in the natural biotopes in order to create an optimal environment for the fishes in the aquarium at home. Over the years, his journeys were planned with more and more precisely defined tasks and specific study-goals, may it be to verify doubtful distribution records or to collect material for the work on taxonomical problems.

He made other aquarists profiting from his experiences by lectures, but especially by publications in both national and international periodicals.

WOLFGANG STAECK, born 1939, studied biology and English literatur at the Freie University Berlin, Germany. After his State Diploma he worked as an associate researcher at the Technische University Berlin for several years. In the year 1972 he conferred a degree with the minor subjects Zoology and Botany.

Dr. STAECK is known to a wide public through numerous lectures and the publication of books and papers in journals. Since 1966 he has published a vast number of contributions on Cichlids in German and foreign magazines. Since his major interest is focused on behavioural studies in Cichlids, he is still an aquarist today and familiar with the maintenance and breeding of Cichlids with the experience of many years.

During his numerous study-trips which were primarily intended to learn more about Cichlids in their natural environments and resulted in the discoveries of many new species, subspecies, and colourvarieties, he travelled East Africa especially, but also West Africa and Madagascar. In recent times he undertook journeys to Central and South America to study and collect Cichlids in Mexico, Brazil, Ecuador, Venezuela, Peru, and Bolivia.

A high priority of his research was spent on the Cichlids of the Lakes Malawi and Tanganyika. Not only in these waters but also in rivers of Central and South America, he observed and took photographs of the world of fishes as a diver. Through this he managed to document the ecology and the inhabited biotopes of many Cichlids for the very first time in underwater photographs. As a result of his study-trips he published scientific descriptions of several new species of Cichlids.